The Substance of Civilization

The Substance

of Civilization

Materials and Human History
from the Stone Age to the Age of Silicon

Stephen L. Sass

ARCADE PUBLISHING • NEW YORK

FIRST EDITION

Library of Congress Cataloging-in-Publication Data

Sass, Stephen L.
    The substance of civilization : materials and human history from
the stone age to the age of silicon / by Stephen L. Sass.
        p.   cm.
    Includes bibliographical references and index
    ISBN 1-55970-371-7 (hc)
    ISBN 1-55970-473-X (pb)
    1. Materials. 2. Technology and civilization. I. Title.
TA403.S335 1998
620.1'1'09—dc20   97-44802

Published in the United States by Arcade Publishing, Inc., New York
Distributed by Time Warner Trade Publishing

10 9 8 7 6 5 4 3 2 1

Designed by API

BP

PRINTED IN THE UNITED STATES OF AMERICA

*To Karen, Adam, and Erik—my family,
and the rock upon which my house is built*

Contents

Acknowledgments   ix
Introduction   1
1   The Ages of Stone and Clay   13
2   The Age of Metals: A Primer   38
3   Copper and Bronze   49
4   Gold, Silver, and the Rise of Empires   68
5   The Age of Iron   82
6   A Quick History of Glass   98
7   Building for the Ages   124
8   Innovations from the East   134
9   Stoking the Furnace of Capitalism   147
10 The Birth of Modern Metals   176
11 Steel: Master of Them All   203
12 Exploding Billiard Balls and Other Polymers   215
13 Diamond: The Superlative Substance   238
14 Composites: The Lesson of Nature   250
15 The Age of Silicon   265
    Epilogue: Materials in the Twenty-First Century   277
    Notes   283
    Bibliography   288

Acknowledgments

I OWE DEBTS of gratitude to many who have contributed to this book, in ways both large and small. The most credit for helping to make it understandable to nonscientists must go to my wife, Karen, to whom fell the thankless task of reading the first draft. She did her best to take the scientist's passive voice and make it more active. My sons, Erik and Adam, read the final draft and added their own deft touches to it. Adam came up with the book's title. My editor at Arcade, Timothy Bent, was always encouraging, and skillful in his attempts to make my text more accessible. I am particularly grateful to my close friend Sally Gregory Kohlstedt, who found me wandering around Cornell's Olin Library one day while looking for books on archaeology, and gently pushed me in the right direction. I also greatly appreciated the early readings of the manuscript by two wonderful materials science and engineering undergraduates, Brian Elliott and Jessika Trancik, both of whom made a number of helpful suggestions. Mae Ryan of Johnson Matthey tracked down information for the chapter on platinum. Finally, I need to thank Cornell for giving an engineering professor the opportunity to teach a course to students in the humanities.

While I gratefully acknowledge the significant

contributions of all of the above, I also bear all final responsibility for the text, and for whatever errors it may contain.

Stephen L. Sass
Ithaca, N.Y.

The Substance of Civilization

Introduction

SPRING HAD COME TO ITHACA — for the second or third time that year — with mild temperatures melting the mounds of grimy snow, snowdrops peeping through here and there, and V's of Canada geese honking exuberantly overhead on their journey northward. I was giving a lecture to my sophomore-level materials science class at Cornell. A glance at the students told me I was losing them in the haze of an April morning. I wondered what I could do to prop open their spring-heavy eyelids. I had been talking about the heat treatment of steel. In an act of desperation and hope, I abandoned my course notes.

"Isn't it remarkable," I asked, "that just a sprinkle of charcoal, which we use in our backyard barbecues, changes iron into steel, and transforms a weak metal into a strong one? And isn't it lucky that both iron and charcoal are so cheap? What form would our world take without iron and steel?"

The change in my voice caused a few eyes to open. One student replied. "Well, it's hard to imagine a Corvette without iron and steel."

"And of course sports cars are the highest expressions of civilization," I teased the student. "In addition to your car," I continued, "our great cities would not exist today. There would be no spectacular bridges, no skyscrapers housing tens of thousands of people. Steel became cheap just after the middle of the nineteenth century, thanks to the ingenuity of the English inventor Henry Bessemer. Before the Bessemer process, train tracks made of wrought iron were quickly squashed out of shape and had to be rotated every three to six months. Imagine doing that in today's subways."

I seemed to have attracted their attention, so I pressed on. "Iron was once more valuable than gold. One of the most important technological revolutions of human history was triggered by the transformation of iron from a rare to a common or working metal. We call the era the Iron Age. Perhaps a good name for the past century would be the Steel Age."

I explained that Britain had been the world's leading producer of steel in the nineteenth century. She had lost her position of dominance at the turn of the century, first to the United States, thanks to the vision of Andrew Carnegie, and then to Germany. If Britain's decline in steelmaking foretold her fading as a world power, what was that telling us about the United States, where several major steel companies have failed in the past decade? The United States Steel Corporation recently dropped "steel" from its name to become USX. But perhaps, I speculated, the time when steel production is the primary gauge of industrial strength has passed. Steel has lost its romantic allure, submerged beneath a stream of exotic new materials flowing out of industrial, government, and university laboratories. Today, silicon is the most sensitive indicator of a nation's economic climate. A "chips barometer" that uses a scale graduated in millions of chips per year has replaced the old "steel barometer" whose scale was measured in millions of tons.

Later that morning, between lectures, I thought about other ways I could have made my point about materials and the progress of nations. I thought about my flight home from Washington the night before. Jet aircraft can fly passengers in comfort at altitudes of 30,000 feet due to the high-strength aluminum alloys in their fuselages and the high-strength nickel alloys in their engines. Alloys are critical because, curiously, pure metals are extremely weak. I sometimes illustrate this point to my classes by asking a particularly frail-looking student to try to bend rods of pure aluminum one inch in diameter across his knee, silently praying that he doesn't destroy his joint in the process. But the aluminum never fails me. And the students are always delighted as one of their own twists the bar into a pretzel shape. So aluminum is too weak to be used by itself for fuselages, which protect passengers from the minus 50 degree Fahrenheit temperatures and air pressures lower than those found atop Mount Everest. Metal alloys can be a thousand times stronger than this aluminum bar.

The road to the jet fuselage was paved by disaster for the reason that early designers did not fully understand the materials with which they were working. The tragic crashes near Rome of the British Comets, the first commercial jets, come first to mind. Built by de Havilland in the early 1950s, the jet's designers had made the windows rectangular, and tiny cracks formed in the vicinity of their corners, where stresses were highest. Each time the plane took off and landed, these cracks

3

lengthened, and eventually the fuselage came apart. Happily, today's jet aircraft do much better. (Still, whenever I board a plane, I compulsively scan the area around the door for small cracks. A colleague at Cornell always checks the plane's manufacturing date.) Once in a great while metal fatigue still causes a disaster, such as when part of the fuselage of an Aloha Airlines plane peeled off over the Pacific in 1988.

The Comet accidents had enormous economic consequences. Britain lost its large lead in the commercial jet aircraft business to the United States, where companies like Boeing began to dominate the industry, and still do. The loss was catastrophic for the British economy and perhaps signaled the end of one of the greatest economic empires of the post-Renaissance.

Materials not only affect the destinies of nations but define the periods within which they rise and fall. Materials and the story of human civilization are intertwined, as the naming of eras after materials — the Stone Age, the Bronze Age, the Iron Age — reminds us. For example, turmoil in the eastern Mediterranean toward the end of the second millennium B.C.E., resulting in shortages of bronze, helped launch the Iron Age and made it easier for the Hebrews to establish themselves in the land of Canaan following their exodus from Egypt. The Jews despaired over ever conquering Canaan, the land their God had promised them, not because of a lack of iron will, but very likely because of a lack of iron. This shortage hindered the Jews in the eleventh century B.C.E., a time of transition between the Bronze and Iron Ages, because the Philistines who ruled Canaan had mastered iron metallurgy, and they had not. The Old Testament is filled with references to the connection between materials and human destiny.

> Now there was no smith to be found throughout all the land of Israel; for the Philistines said, "Lest the Hebrews make themselves swords or spear," but every one of the Israelites went down to the Philistines to sharpen his plowshare, his mattock, his ax, or his sickle. . . . So on the day of battle there was neither sword nor spear found in the hand of any of the people with Saul and Jonathan; but Saul and Jonathan had them. (1 Samuel 13:19–22)

Despite these handicaps, Saul was able to defeat the Philistines, at least to a limited extent, before losing favor with God. Saul's victories (including one resulting from a particularly well-slung stone — a throwback, literally and metaphorically — by David) apparently allowed the Israelites to acquire the knowledge of working iron, though the Old Testament does not tell us how this occurred. This new technology may have contributed to the later successes of David and Solomon.

Materials guided the course of history. Iron contributed to the conquest of Canaan first by Sargon of Assyria and then by Nebuchadnezzar of Babylon, and brought about the destruction of Jerusalem and the Babylonian exile of the Jews in the sixth century B.C.E. But the capture of Babylon by the Persian king Cyrus also freed the Jews to return to Palestine, an event that stimulated the development of hotter kilns. Hotter kilns in turn led to the blowing of glass, which transformed glassware — bottles in particular — from rare to commonplace items, and gave the world its first transparent and sturdy windows. Athenian silver mines allowed the Greeks to block the Persians from expanding into the Aegean; gold from Thrace gave Alexander the wherewithal to create an empire such as the world had never seen; China was the birthplace of paper and gunpowder, both of which shaped the modern world it is today struggling to enter.

History is an alloy of all the materials that we have invented or discovered, manipulated, used, and abused, and each has its tale to tell. The stories of some materials, like diamonds and gold and platinum, involve opulence and mystery. The stories of iron and rubber are more mundane, reflecting the fact that they are industrial, indecorous substances. But all these materials have had a profound influence on human history, and to tell the story of each one means spanning many centuries and crossing enormous geographical areas, from South America — source of platinum and rubber, and the great quantities of gold and silver that supported Spain's adventures and misadventures beginning in the sixteenth century — to Great Britain, where the very modern problem of shortages of natural resources triggered a sequence of events leading to the Industrial Revolution, and finally, to the United States, center of material innovation for much of this century, home to the computer and information revolutions ushered in by silicon and optical fibers.

Yet wherever the story of materials takes us, it begins with and returns to the unique properties of each substance. Why, for example, glass shatters when dropped while metal does not. Why rubber is so different from either glass or metal, *except* when it is cold (a fact tragically apparent on a frigid January morning in 1986, when the *Challenger* space shuttle exploded). Why atomic-scale defects make metals weak, and why we can use those defects to make them strong again. Why flaws in ceramics are both different in nature and larger in scale, and why, until these flaws can be controlled, ceramics will never be used in the demanding applications now on the drawing board for the next century, such as the hypersonic plane, capable of taking travelers halfway around the world in a few hours. Why there is a thing such as metal fatigue, the culprit in the Comet and Aloha Airlines accidents. And why the similarity between bone and wood, nature's building materials, and fiberglass and graphite composites has led to the latter's use in boat hulls, fishing poles, and tennis rackets, as well as in *Voyager*, the lightweight airplane that circled the earth without refueling in 1986.

Because materials and their uses have evolved, they lead us back to the foundations of human society, and map the movement from a hunter-gatherer style of life toward a more sedentary existence centered around cities. Dense areas of population develop as the materials that foster them become more sophisticated; the denser the population, the more sophisticated the building blocks. So, too, the higher we go literally (airplanes, skyscrapers) the more complex the substances that take us there.

This book will seek to address the question, "How did materials shape our culture?" It is an enormous question, and there is not one answer but many, as interrelated as a set of Russian nesting dolls. What I do as a materials scientist is part of a continuum stretching back to the beginnings of human life on earth. The urge to innovate, to seek improvements in our material environment, to profit from those improvements, and most important, to survive, have their origins millions of years ago. Early humans were by necessity materials scientists all, constantly testing and improving upon what they found at hand. A comparison of what they had to work with and what we have today illuminates what our lives would be like were a particular substance not available to us.

What the earliest humans found at hand was stone. Later they discovered clay and how to fire it, which was hugely important to the growth and viability of large cities. This innovation (leading to the production of ceramic pots in large quantities), which occurred in the area of Anatolia, in modern-day Turkey, eight thousand years ago, made possible the easy cooking and storage of liquids and grains, as well as their transport. Since the earliest known examples of writing appear on clay tablets unearthed in Mesopotamia, we know that clay was as important for storing information as it was for storing food. Because of the durability of these tablets, archaeologists have been able to recover remarkably detailed records going back five thousand years. In fact, clay tablets had a far greater chance of surviving over the millennia than papyrus, so we have more "hard" facts about life in ancient Mesopotamia than we do about events that occurred in Palestine a thousand years later. Clay is still vital for storage, since one of its elemental constituents is silicon, whose semiconducting properties form the basis of the personal computers that today store most of our data.

Few countries would be able to survive solely by exporting their natural resources or agricultural products. The economic security of most nations has always depended on their ability to manufacture and market technologically advanced products. With limited natural resources and farms hopelessly unsuited to competing in the international marketplace, Japan, until recently, provided a textbook example of a country able to sustain a robust economy almost entirely through technological capabilities. This means filling a continuous demand for improvements on old materials, as well as inventing wholly new ones. Innovations often occur when people are experimenting with new ways to process old materials. As Tadahiro Sekimoto, until recently president of Nippon Electric Corporation, said, "Those who dominate materials, dominate technology." And, the ancient Romans might have added, dominate the world.

Today, substances such as silicon are revolutionizing the way we live. Others, while still laboratory curiosities, are emerging. What materials await discovery? Perhaps some that will enable hypersonic flight at twenty-five times the speed of sound. They may seem as fantastical as the "tritanium" and "dilithium crystals" so beloved by aficionados of *Star Trek*, but they are waiting to be found. We are currently witnessing an intense international effort involving high-temperature superconductors, which, below a particular temperature, have absolutely no resistance to electrical current. This property holds out the promise of a dazzling variety of innovations, including high-speed levitating trains and electricity transmission without energy loss. But despite much hyperbole, at present these superconductors are making profits primarily for companies marketing demonstration kits for educational purposes and supplying materials to research laboratories. Materials scientists will have to study these superconductors for many more years before they find significant commercial applications. Moreover, these applications will likely emerge serendipitously — by accident and luck — from research projects whose initial goals were quite different.

This, too, has always been the case. As it was with their ancient counterparts, what materials scientists discover is most often not what they had started out looking for. And what motivated early humans still drives us. Whether necessity, greed, or an unstoppable curiosity, these motivations, when we can identify and understand them, might give us the wisdom to avoid the mistakes of the past while matching its greatest successes.

I have a further reason for writing this book. My children have grown up in an affluent society, surrounded by and benefiting from technologies that make their lives secure and comfortable. They fly across the country in just a few hours to visit their grandparents in Idaho. My oldest son, Adam, talks on the telephone with a friend in Israel seconds after direct dialing. When I was a high school student in New York City in the 1950s, I did my complex mathematical calculations on a slide rule that swung rakishly from my hip, calculations that my son Erik now does on his ten-dollar, credit card–sized calculator in a fraction of the time and with far greater precision. My children have very little idea of what is behind these and other marvelous inventions, which they see as so commonplace. This book is to help them appreciate and wonder at the material nature of our world. Perhaps with talent and luck and perseverance they or their peers will discover an extraordinary new substance.

Inherent in my tale, I hope, is also the excitement of discovery. Trying to convey this very special feeling to students in my laboratory, or in answer to my wife Karen's question, "why do you enjoy doing research?," I often relate the story of the unearthing of the tomb of Tutankhamen in 1922 by the archaeologist Howard Carter. "Can you see anything?" asked his patron, Lord Carnarvon, who was standing anxiously behind Carter as he peered into the tomb of Tutankhamen for the first time. "Yes," Carter replied, "wonderful things!"

A disclaimer is in order. As a materials scientist concerned with the relationship between metals and ceramics, my research involves manipulating these materials' atomic structure and microstructure, to formulate new substances. I am neither an archaeologist nor an historian. My primary hope here is to offer an overview of the synthesis of materials and history. The authors of many archaeological and historical surveys may recognize my debt to them, and it is a debt I gratefully acknowledge.

Since the time scales involved span thousands and, in some cases, millions of years, the following approximate chronology provides a frame of reference for the human activities discussed here:

The Substance of Civilization

Historical Name	Time Period B.C.E.
Stone Age	2,000,000
Lower Paleolithic	1,500,000
Upper Paleolithic	40,000
Neolithic	8500
Modern Era (End of the Ice Age)	8000
Chalcolithic	4500
Bronze Age	3150
Abraham, Isaac, and Jacob	1750
Iron Age	1200
Moses and the Exodus	1200
Alexander the Great	330
Caesar Augustus rules Rome	0

Unless noted otherwise, all temperatures in the text are in Celsius.

1

The Ages of Stone and Clay

> The High Priest of Kulaba formed some clay and
> wrote words on it as if on a tablet —
> In those days words written on clay tablets did not
> exist,
> But now, with the sun's rising, so it was!
> The High Priest of Kulaba wrote words as if on a
> tablet, and so it was!
>
> — Enmerkar and the Lord of Arratá[1]

GAZING ACROSS THE STARK, sunbaked land- and wa-
terscape of Salmon Creek Reservoir, set in the sagebrush
desert of southern Idaho, I was alert to any motion of
the tip of my fishing pole, propped up by rocks. My
family and I often visit my in-laws in Twin Falls, Idaho,
and we always go fishing for rainbow trout. Erik, my
younger son, back from exploring the barren cliffs,
came running up to me, clutching a black stone different

in appearance from the slabs of lava rock scattered along the shore about us.

"Dad, what's this?"

Turning the dull stone over in my hand, I told him it was obsidian. "It's glass — different from most rocks. More like a frozen liquid than a crystalline solid." When I started to explain that it had a different atomic structure than many other minerals, his gaze drifted away. I turned and threw the piece of obsidian against a nearby rock, shattering it into shiny, razor-edged chunks. "Native Americans around here and people in the ancient Near East used obsidian to make axes and arrowheads, because it splits into lots of sharp pieces." Glass is one of many materials that craftspeople used thousands of years ago that we still employ, albeit in very different ways. I told Erik that today phone companies were replacing copper wires with optical fibers made of very pure and ultra-clear glass.

"Well," Erik asked, perhaps less concerned with these facts than the rock he had found, "why is obsidian black?"

"Clear glass is made from silicon, oxygen, sodium, and calcium," I replied. "But obsidian contains dirt, small amounts of other atoms that make it black. The first people could make tools out of rocks like this, which is why we humans did so well. Glass was as high tech ten thousand years ago as it is today." Satisfied, Erik checked his rod and went off to look for other rocks.

A few days later, we were looking out across a large moraine at the spectacular vistas in Rocky Mountain

National Park. Once the basin before us was clogged with glacial debris — large boulders and rocks — now hardly to be seen, though the U-shaped valley is the signature of a glacier melted long ago. The displays in the park exhibit at Moraine Basin remind visitors that rocks erode because slightly acidic water attacks the cement that holds minerals together. Most rocks do not have a uniform structure like obsidian, but are composites of several different constituents, similar to concrete, an artificial pourable stone, in which mortar bonds together sand and hard rocks. In nature, heavy loads are always supported by composite structures, not homogeneous materials. Mimicking nature, humans also use composite materials for their most advanced applications. We'll learn why later. First let's turn to the earliest tales of materials.

Early humans faced overwhelming obstacles to survival. They needed food for sustenance, weapons against predators — both animal and human — and shelter from an often brutal environment. In their desperate struggle, our ancestors came to realize that the gray-brown-black rocks they found scattered about them were useful for making weapons and tools; flint and obsidian were particularly desirable. Anthropologists have uncovered the earliest evidence of stone implements in the Rift Valley of East Africa. More than two million years ago, humans were first finding ways to master nature, and the earliest stone artifacts discovered in the Olduvai Gorge consisted of flakes, or thin chips, and the stones from which they were struck. No one yet knows what these stone implements were used for, although the fact they were frequently found near bone fragments suggests that our ancestors used them to butcher animals, ranging in size from elephants and hippopotami to rodents and tortoises. Sharp flakes could slice through tough hides, while stones broke open bones to get at the marrow. Since anthropologists believe the diet of these early folk was more than half vegetarian, they likely also used stone tools to dig up roots and tubers, and crack open hard-shelled nuts.

Perhaps not coincidentally, the separation of the genus *Homo* (man) from *Australopithecus* (southern ape) occurred at approximately the same time as the first appearance of simple stone tools. It is still unclear whether *Homo* fashioned rudimentary tools and *Australopithecus* did not; it is conceivable, however, that the ability to make tools gave *Homo* an advantage in the battle for survival over *Australopithecus*, which eventually became extinct. Nevertheless, *Homo* survived for other reasons, including the sharing of responsibility for gathering wild plants and hunting game. Cooperation within a hunter-gatherer group was the first step toward the specialization in crafts leading to innovations in both their world and, ultimately, ours.

Early development of simple collections of stone implements — or tool kits, as they are called — appears to have taken place entirely in Africa and progressed at a very slow pace, because food was readily available. Where there was enough to eat, there was no need to innovate. Pressure on the food supply by an increasing population forced these early folk to develop new and better ways to hunt and gather food. It was during the Lower

Paleolithic period that early humans spread out of Africa to parts of Asia and Europe, perhaps 1.5 million years ago. At about this time, the ancestor of modern humans, the species *Homo erectus* (man who walks upright) emerged. *Homo erectus,* with a brain larger than such predecessors as *Australopithecus,* walked with a striding gait and differed from modern humans primarily through its larger jaws and teeth. With additional enlargement of its brain and changes in facial structure, a new species, *Homo sapiens* (wise man), appeared around 250,000 years ago. Further division into subspecies occurred, including the famed *Homo sapiens neanderthalensis* and culminating in *Homo sapiens sapiens,* modern humans who emerged 100,000 years ago, and who replaced all other human types on earth by approximately 30,000 years ago.

Among our ancestors' tools, hand axes — large cutting implements with two planar faces meeting at a shallow angle — were the primary product of stoneworking, though early peoples also fashioned scrapers for cleaning animal hides, as well as knives and toothed implements. With time, artisans developed sophisticated flaking techniques, which they used masterfully to craft flint tools. Instead of laboriously chipping individual blades, they first carefully prepared a plump cigar-shaped flint core, from which they then rapidly struck several blades — an early version of mass production. As the final step in this "Levallois technique," as it is called, stoneworkers retouched the edges of the flint blades with well-aimed blows.

Cementing hand axes into wooden handles with tree resin or bitumen (allowing ingenious early artisans to take advantage of the lever principle and increase the velocity of the axblow), a crucial innovation. They fashioned bone, ivory, and antler into new tools. One such class of new implement was the straightener for wooden shafts. These straighteners enabled the invention of the arrow, which revolutionized both hunting and fighting by allowing killing at a distance. By this time, high-quality stones were much in demand and were frequently traded over long distances. Flint, for example, was shipped eighty miles to sites in the former Soviet Union.

To survive the harsh winters of middle and northern Europe, *Homo sapiens* stitched together animal skins using bone needles. Finds from north of Moscow, dating from 22,000 years ago, include leather caps, shirts, jackets, trousers, and moccasins. Fox and wolf furs provided additional insulation. These people also built substantial shelters, fashioning the walls of their huts from large numbers of mammoth bones. They made stone floors, lamps holding animal fat to light the long winter nights, and fireplaces for heat and cooking. Excavations show they understood the role of draft in making hotter fires, for their hearths were frequently complex structures, with corrugated floors to increase the flow of air to the wood fuel. Many thousands of years later, this simple idea would lead to furnaces hot enough first to fire clay pieces, then to extract copper and iron, and finally, to melt and blow glass.

The stone tools in use just before the start of the Holocene, or Modern Era, usually dated ten thousand years ago, indicate that humans were still hunter-gatherers. Archaeologists have unearthed spear and arrow tips fabricated from both stone and bone, with long grooves to make wounded animals bleed more heavily, thereby speeding their death. Together with wooden shafts found in northern Germany, they demonstrate that hunters had added bows and arrows to their arsenal. Fishhooks made of bone originate from the same period, as do weapons such as the harpoon and the spear thrower, a leather-thonged sling, allowing hunters to increase the range and velocity, and therefore the impact of their spears, encouraging them to stalk larger and more dangerous game. Mammals that had survived earlier ice ages, such as the woolly mammoth and great deer in Europe, and the giant buffalo and giant Cape horse in Africa, became extinct toward the end of the last Ice Age, possibly because of the increased sophistication of the hunters and their weapons.

An extraordinary discovery occurred roughly 26,000 years ago, when artisans at a site in what is now the Czech Republic mixed clay with loess, rich soil left behind by retreating glaciers, and then fired it in ovens (the word *ceramic* comes from the Greek *keramos,* meaning "burnt stuff"). Clay was the first substance that humans totally transformed by heating. Many modern materials undergo similar metamorphoses when heated. Soft doughy clay is remarkable because in a kiln or oven it becomes a hard, heat-resistant ceramic that can hold liquids. Along with the discovery that grains could be cultivated, the advent of ceramics led to what some have called the Neolithic ("new stone") Revolution, laying the agricultural basis for the first cities of the world that emerged in Mesopotamia, between the Euphrates and Tigris Rivers in what is now Iraq.

Artisans at the Czech site fashioned Venus-like figurines from clay. Curiously, most of these were found shattered, and scientists speculate that they were fired so as to break purposely, perhaps as part of a religious ritual whose meanings are lost to us. Ceramic pots were first fired in Japan twelve thousand years ago, but only much later, during the seventh millennium B.C.E., were potters able to make them in sufficient quantities to have any significant impact on the economy of early villages.

Discovering how to produce a ceramic by firing clay was a milestone in our quest to master nature, and its importance can't be overstated. Up to that time craftspeople had transformed stone, bone, and wood into tools and weapons by altering their shape, not their intrinsic properties. With ceramics from clay, humans learned that the physical properties of materials could be dramatically improved.

In antiquity, most advances involving materials were based in some manner on this simple concept. (They frequently are today as well.) Early humans had discovered that heating flint and chert allowed them to be more easily cleaved, so, strictly speaking, clay was not the first material humans modified. Clay is, however, the earliest example of a complete change in the property of a material that involved a major technological advance; it was, in this sense, the first truly man-made material.

Clay had the additional advantage of being mal-

leable, making it easier to use than stone in the fashioning of pots and sickles, both rather complex objects.
Clay can be easily shaped because its layered atomic
structure leads to the formation of thin plates called
lamellae, each weakly bonded to its neighbor. Individual
layers are built up by periodically repeating an octahedral, or eight-sided, arrangement of oxygen atoms en-
closing an aluminum atom (imagine an eight-sided cage
of oxygen atoms, with a triangular array of three oxygen atoms making up each side, and with an aluminum
atom at the center of the cage), and a tetrahedral, or
four-sided, arrangement of oxygen atoms with a silicon
atom at its center (imagine a four-sided cage of oxygen
atoms, with a triangular array of three oxygen atoms
making up each side, with a silicon atom at the center of
the cage). Both the octahedra and tetrahedra are tilted
over on their triangular sides. Larger platelike clay crystals are built up by interweaving layers of octahedra
with layers of tetrahedra and water molecules. These
thin crystals move easily past one another, like playing
cards in a deck, allowing clay to be shaped. Clay's remarkable transformation to a ceramic occurs when water and hydroxyl molecules (made up of a hydrogen and
an oxygen atom) are driven off upon heating. As the
water evaporates, clay shrinks significantly, and new
atomic structures are created.

When the atomic structure of clay changes, so do its
properties. The element carbon demonstrates this change
even more dramatically. As graphite, which has a layered
structure similar to clay, carbon is slippery, soft, and
black, hence its use as a dry lubricant and as the lead in

pencils. Under extremely high pressures and temperatures — found in nature only deep within the earth — graphite transforms into diamond, which has quite a different atomic arrangement: a three-dimensional network instead of a two-dimensional, layered structure. This network makes diamond transparent and extremely hard; in fact, it is the hardest substance known on earth.

How was the firing of clay first discovered? Perhaps it was by a mason observing the effect of the sun on his bricks. Or by a woman idly shaping clay during cooking, dropping her handiwork into the campfire and then finding a rock-hard solid among the ashes. From such mundane events, lost to recorded history, are born revolutions in technology.

Once our ancestors had learned how to transform clay into ceramic, they applied this valuable lesson over and over again, developing new ways to manipulate materials' properties to their benefit by using both heat and mechanical energy (in other words, hammering). Sometimes the structure would be changed on the atomic level, as when clay is fired to a ceramic, and sometimes on a larger scale, as when metals are forged into new shapes. While the process of firing clay to obtain ceramics is commonplace (schoolchildren do it every day), its discovery made possible innumerable innovations in materials.

Most historians consider the end of the last glacial period, about ten thousand years ago, as the start of the Modern Era. As the average temperature on earth increased, ice fields receded and oceans rose from their lowest point of 400 feet below today's levels, engulfing vast stretches of coastline and severing the land bridge between Asia and North America. Loess deposited by retreating glaciers became fertile ground for meadows and pioneering trees such as birches, creating verdant pastures for animals and a rich bounty for hunters. Humans still made tools and weapons from stone, and so the beginnings of the Modern Era overlap with the end of the Stone Age (defined as the Neolithic and Chalcolithic — "copper stone" — Periods). Of course, different civilizations begin using particular materials at different times; indeed, some native tribes in the Amazon still use stone implements. For my purposes, I will define the start of an age as the moment when a material first began playing a significant role in the life of a society.

The transition from a nomadic hunter-gatherer lifestyle to a sedentary existence was crucial and first occurred, so far as we know, in the Near East. Evidence for this gradual shift is found in the Natufian culture (10,300 to 8500 B.C.E), which derives its name from Wadi en Natuf, *wadi* being Arabic for "streambed," where traces of its existence were first uncovered in the environs of present-day Israel. The Natufians established permanent settlements, sometimes exceeding one hundred inhabitants. These settlements typically overlooked marshy areas, which attracted game in search of water and provided ample opportunities for hunting, as well as for fishing and gathering vegetables. Their front yard was their larder. Caves overlooking the sea at

Kebara at Mount Carmel and an open-air site facing the Jordan River at Eynan in the Hula Valley were all hospitable locations.

Archaeologists investigated three levels of villages at Eynan. Each consists of fifty round houses built of stone, with floors below ground level, walls three feet high, and roofs of conical or hemispherical shape. Since there was no evidence for either the domestication of animals other than dogs or the cultivation of grain, their inhabitants must have still lived by hunting and gathering. Enormous quantities of gazelle bones were found in caves at Mount Carmel, indicating that gazelles were the primary source of meat. The discovery of large numbers of flint sickle blades and grinding utensils, including large stone mortars, provides strong evidence that Natufians gathered wild grains. Sickle blades with edges that show considerable wear are occasionally found mounted in beautifully carved bone handles. Relatively easy access to food gave the Natufians spare time to lavish rich artistry on their implements. They also developed new tools for hunting and fishing, including harpoons, hooks, and net sinkers. Highly prized obsidian was not present in Palestine and Syria, so artisans fashioned their stone tools and weapons out of local materials.

Natufians were able to feed themselves in relatively large numbers by hunting and gathering while still maintaining a permanent settlement. But evidence suggests that they were also on the brink of a variety of innovations, including the cultivation of grains and the domestication of animals. When those finally occurred, they were followed by the rapid growth of permanent towns.

Jericho, situated on an oasis 820 feet below sea level in the lower Jordan River Valley, just north of the Dead Sea, is a fine example of such an early agricultural settlement. Today, all that remains of this ancient community is a 70-foot-high mound about ten acres in area, thoroughly excavated by archaeologists. Jericho owes its existence to a perennial stream in the oasis, which allows extensive agriculture. Thus, eight thousand years before the time of Christ, Jericho supported a population of at least two thousand. They could feed themselves, thanks to their agricultural practices and ever-present water. But what brought people to Jericho was its access to salt, sulfur, and bitumen, all valuable minerals taken from the nearby Dead Sea.

Archaeologists at Jericho have unearthed a burned Natufian shrine, which was replaced first by a village and then a town occupying at least ten acres. Its inhabitants built round houses of sunbaked mud bricks on a stone foundation with floors well below ground level. The curious loaf shape of these bricks made them very inefficient for construction. Initially Jericho had no fortifications, but as it grew wealthy, city walls were needed for security. A rock-cut ditch, twenty-nine feet wide and seven feet deep, was dug without the help of picks — an incredible feat, considering that ten thousand years ago stone tools were used to cut stone. Within the ditch, workers built and rebuilt a stone wall five feet thick as rubble gathered around its base. They also erected a large circular stone tower with an internal staircase, which, after nearly ten millennia, still stands to a height of just over twenty-six feet. Water channels ran from the top of the tower down to a cistern at its base. Constructing these defenses required both a large population and a central organization for planning and finance.

To grow the large amounts of grains required to sustain such numbers, Jericho's inhabitants needed to domesticate wheat, craft tools, till the soil, harvest and grind the wheat, and, finally, find ways to store their harvests. By 8000 B.C.E., fully agricultural villages were established in Syria and Palestine, supporting much larger populations than in other parts of the Near East.

Archaeologists uncovered evidence that early farmers, taking advantage of the relatively abundant ground water, transplanted wild grains from the neighboring Judaean hills and cultivated them in the environs of Jericho. A major step forward in the establishment of an agriculturally based economy was the development of wheat that could be harvested with a sickle. Wild cereals have heads that shatter when their seeds ripen, since this is nature's way of sowing the next year's crops. An inability to sow their own seeds would make it difficult for plants to reproduce and would eventually lead to their extinction, barring intervention of another seed-spreading mechanism such as birds. Initially, village folk harvested wild cereals in the Judaean hills by tapping the plant stems and catching the falling seeds in a sack. Due to mutations, however, some of the wild cereals had heads that did not shatter. If food gatherers tried to harvest wheat with a sickle, the seeds from the shattering heads were lost on the ground, while the seeds from the unshattered heads were collected. By planting this type of wheat apart from the dominant wild wheat, farmers at Jericho cultivated crops with non-shattering heads, which could be efficiently harvested with sickles. Hybrids of these early types of wheat are still used today.

As wheat domestication became more widespread, farmers and merchants needed better ways to store their foodstuffs than in containers sewn from skins or woven from straw, or in holes in the ground. Stone vessels of limestone, marble, or quartz were hardly suitable for storing large quantities of food or daily cooking because they were so difficult to fashion; those that did exist were likely reserved for religious rituals. Toward the end of the seventh millennium B.C.E., artisans began modeling slabs of clay into pots and baking them in open fires, creating for the first time an ample supply of containers. Large-scale storage of liquids and solids was now feasible. Though fragile, ceramic pots were also easy to make. (Archaeologists from the beginning of this century frequently found ancient sites littered with large quantities of potsherds, which were used to assign dates to different levels at their digs, until other methods such as carbon-14 dating were developed. Carbon dating, however, depends on finding artifacts containing organic material, which, unfortunately, decays readily.) Farmers could now store large quantities of grain, protecting them from scavenging animals. Jericho's merchants traded with Mesopotamia and Anatolia, where they bartered for obsidian, malachite, turquoise — so named because it was imported through Turkey — and copper. Traders were exchanging grain for obsidian as early as 8300 B.C.E.

Human affairs rarely proceed at a uniform rate. Typically there is a long quiescent period, followed by a sudden cascade of changes. So it was with the Neolithic Revolution in the Near East, where expansion of the area occupied by agricultural villages increased the opportunities for interaction between them. The ferment that resulted spurred further innovations in farming, continuing the cycle of growth. Major developments in the seventh millennium B.C.E. included large-scale production of ceramic pots, as mentioned, the proliferation of sheep as an important domestic animal, the spread of villages to islands and coastal highlands, and finally the emergence of large towns. The center of progress shifted north, from Jericho and its surroundings to Anatolia, where the city of Çatal Hüyük is located. The first appearance of clay containers at Çatal Hüyük took place in approximately 6500 B.C.E.

Located in a well-watered region on the Konya plain, Çatal Hüyük has provided us with an enormous amount of information about the activities of a large town of the Neolithic period. A river flowing past the town, fed by the Taurus Mountains, allowed farmers to use simple irrigation methods. Their successes likely grew out of agricultural practices originating in Jericho, practices they learned about through trade.

At its height, with more than one thousand single-story houses of sunbaked bricks, Çatal Hüyük boasted five thousand inhabitants. Houses were arrayed in terraces on the side of a hill. There were no streets; people entered and exited their homes through the roof, using ladders. Solid outer walls gave protection from marauding nomads. Each house had a single rectangular room with an area of about 265 square feet, roughly the size of a small studio apartment in Manhattan.

Çatal Hüyük's economy was based on agriculture, stock breeding, and the hunting of wild cattle. Domesticating the auroch, a giant wild cow with horns as long as seven feet, must have sorely tested the bravery of the townspeople. The remains of a man gored in the groin were excavated from a burial site there. The inhabitants worshiped the auroch, testifying to its centrality to their lives, and among the most impressive, unique archaeological discoveries at Çatal Hüyük are shrines to these wild cattle.

Active volcanoes in its neighborhood gave Çatal Hüyük a nearly total monopoly in obsidian. Townsfolk traded it for copper, lead, and seashells, as well as flint from Syria for fire making and a variety of tools. As early as 6400 B.C.E., artisans worked metals into beads, pendants, and tubes. Potters also fashioned pots from slabs or coils of clay, smoothing their surfaces with a paddle and anvil. This was a slow process. Fifteen hundred years would pass before the pottery wheel was invented, sometime at the end of the fifth or the beginning of the fourth millennium.

By the beginning of the sixth millennium B.C.E., Çatal Hüyük's importance had faded away and the center of technological innovation had once more shifted, this time to Mesopotamia (present-day Iraq), where large urban civilizations developed. We might reasonably wonder how this region, which had made no contributions to the early advances in the Near East, and which had few natural resources, became the site of the world's first large cities. Again the answer lies, to a large degree, in its inhabitants' use of materials.

The early locations of urban development in Mesopotamia were situated in the most fertile areas — the alluvial plains between the Tigris and the Euphrates, and the moist portions of the northern plains. To the north and east, on the borders of Turkey and Persia (Iran), the mountains rise to 10,000 feet. To the west of the Euphrates lie the Syrian and Arabian Deserts. In flood, the Tigris and Euphrates carry silt, depositing an alluvial plain extending from 100 miles north of Baghdad down to the Persian Gulf. Given adequate water, this plain of rich soil can be quite productive. The river deltas in those days were at least 95 miles farther north of where they are today, though arguments continue over exactly how much.

Mesopotamia has long been divided into two principal parts: the northern, Assyria, and the southern, Babylonia. Ur, Uruk, and Eridu, the earliest cities there, were located along the meandering course of the Euphrates in Sumer, in southern Babylonia, where irrigation was made possible by the level terrain (the Euphrates drops only 160 feet on its 500-mile journey through the Mesopotamian plain to the Gulf of Arabia). Large canals branched off the river and from these flowed secondary canals. Still smaller irrigation ditches diverged off these canals, until a web of water spanned the entire countryside. Aerial photography today gives remarkable documentation of the complex networks crisscrossing Sumer, providing irrigation, flood control, and communication, several thousand years before the time of Christ. Very few early cities were located on the Tigris, however, because its course was straighter and its current stronger than the Euphrates, making it more difficult for farmers to dig irrigation canals through its banks.

And so southern Mesopotamia became the location of the world's first large cities. Besides the fertility of the soil and the plentiful sunlight, farmers there collaborated to make irrigation succeed on a large scale, and this created an agricultural system such as the world had never before seen. Sumerian temples, the latter-day manifestation of the desert caches, or treasuries, of earlier nomadic tribes, came to play a central role in secular affairs as much as religious. They acted as clearinghouses for farmers and supported artisans with their patronage. Temple officials also coordinated the upkeep of the irrigation systems, facilitating the fair distribution of water and the cooperation needed to regularly clear canals and ditches of heavy river silt and rebuild dikes.

While we tend to think of trade as a relatively modern phenomenon, it flourished in southern Mesopotamia during the fourth millennium B.C.E. Once Sumerians had perfected techniques for growing food in large quantities under relatively stable conditions, they traded their surplus to bring the products of the world to their cities. With the exception of grain, and the mud that they used for construction, Sumerians imported almost everything else they needed, including wood, stone, and copper. By the close of the fourth millennium, they controlled the long-distance trade routes supplying these raw materials. There were Sumerian merchant colonies or provincial administrative centers located in Syria and Iran, some as far away as 500 miles from southern Mesopotamia. Seals on clay tablets unearthed in a large administrative building at Jamdat Nasr suggest the existence of a league of Sumerian city-states, the forerunner of a centralized form of government.

Once a stable supply of food was assured, specialization flourished. Artisans were free to devote themselves fully to their vocation. And with specialization came the potential for innovation, since gifted craftspeople — the forebears of the materials scientist — had both the time and encouragement to experiment with and improve on their processes and their handiwork.

One other development that was crucial to the flourishing of Sumerian society was written communication. Writing enabled temple officials to keep track of the vast quantities of cattle, sheep, grain, and wool entering and leaving their storehouses. It helped Sumerian merchants to keep accounts of their trade with distant lands, and it helped Sumerian administrators to maintain the property deeds that now were mandatory because irrigation had made certain fields more valuable than others. Records based on human memory are inadequate in a complex society, and the need to store information and communicate over long distances made the development of a written language an urgent priority.

The prodigious leap to a written language was made in Sumer at the close of the fourth millennium, most likely for the mundane task of keeping track of farm goods. Clay tablets containing pictographs — pictures representing objects — were unearthed at Uruk and provide the earliest evidence of written language. Thus clay, so critical for storing food, came to play an equally important role in storing information late in the fourth millennium B.C.E. The symbols, drawn with a reed or pointed stick, gradually evolved into stylized pictures. Since the scribes' role was to communicate rather than to create great art, they standardized the pictures into a few stylized marks for simplicity's sake. As the symbols moved away from representations of objects, it was no longer necessary to use a finely tipped drawing stylus; the blunt back end served just as well. Because of this change in style, symbols using groups of wedge-shaped signs evolved, and so was born cuneiform writing (from the Latin *cuneus*, meaning "wedge"). I should add that it is possible writing was invented before this time. Earlier scribes might have written on material such as wood, which did not survive over the millennia to be excavated. That the scribes of 3000 B.C.E. in Mesopotamia used clay and not a more perishable material such as papyrus was fortunate indeed for archaeologists, since clay tablets have a far greater chance of survival than papyrus.

Development of a written language had profound consequences, since not only could records be kept but the hard-won knowledge gained by one generation could be passed on to the next. With the ability to record, retrieve, and transport accurate information, a centralized government could now control distant colonies. The recording of facts allowed information to be compared and manipulated — critical to the development of science. And of course, written language gave birth to literature. Possibly the earliest example is *The Epic of Gilgamesh*, the saga of a king of Uruk from the third millennium B.C.E., which includes the tale of a flood that may have been the origin of the biblical story of Noah. Impressed in clay at the beginning of the second millennium and relating a much older tale, *Gilgamesh* predates Homer and the historical part of the Bible. It is impossible to imagine what life would be like without the results of a simple act of genius: the application of a stylus to wet clay five thousand years ago in a hot dusty city in Mesopotamia.

Archaeologists have recovered vast libraries of clay tablets, providing insight into the activities of life, great and small, in Mesopotamia. A law inscribed around 2000 B.C.E. required that all transactions be recorded, and, since most people were illiterate, scribes were hired to record deals between merchants. After the details of the transaction were inscribed on a clay tablet, the parties applied cylinder seals to it, then the scribe enclosed the tablet in a clay envelope on whose outer surface the contents of the tablet were repeated. If a disagreement arose, the litigants could obtain an unaltered record of their agreement by breaking open the envelope before a legal authority.

In addition to Ur, Uruk, and Eridu, the city of Lagash, situated farther east across the Euphrates and the Tigris Rivers, also flourished. A great deal is known about its ruler, Gudea, who excelled at having descriptions of his achievements inscribed on statues of himself, which now reside in museums around the world. Cuneiform engravings tell us much about both the materials that Gudea sought for his temple to the god Ningirsu and trade at the close of the third millennium B.C.E.:

> Cedar beams from the Cedar mountain [Lebanon]
> He had landed on the quayside. . . .
> Gudea had . . . bitumen and gypsum
> Brought in . . . ships from the hills of Madga
> [Kirkuk?]. . . .
> Gold dust was brought to the city-ruler from the
> Gold land [Armenia]. . . .
> Shining precious metal came to Gudea from abroad,
> Bright carnelian came from Melukhkha [the Indus
> Valley].[2]

Living in a flat and seemingly endless plain, Gudea and Sargon, ruler of Agade, both viewed the sources of their raw materials as mountains made of cedar, silver, and lapis lazuli, some of which were real and others products of their imagination.

The history of Mesopotamia came to involve a *lugal*, or "big man," chosen to lead in times of crisis, eventually becoming a hereditary king. When rulers amassed power, they were challenged and frequently deposed by migratory intruders, who in turn were also overthrown. And so it went, with kings rising and falling, until the eighteenth century B.C.E., when a powerful ruler named Hammurabi emerged from Babylonia and unified much of Mesopotamia under his famous code of law. That code, defining the responsibilities of both king and subject, was discovered inscribed in cuneiform on stone columns. Following Hammurabi's death, the southern part of Babylonia broke away from his empire, and in 1595 B.C.E. Hittites from central Anatolia pillaged the area. By the fourteenth century B.C.E., Assyria, in the northern plains of Mesopotamia, emerged as one of the great powers of the Near East, dominating that region until the close of the seventh century B.C.E., with only occasional interruptions of their control.

Remarkable insight into everyday life as well as events in the halls of power often emerges from cuneiform tablets. In the northwestern Mesopotamian city of Mari, a library of twenty thousand tablets was excavated. They had been part of the palace archives and portrayed life in the city before its conquest by Hammurabi. Located on the Euphrates, the state of Mari was largely agricultural, although its capital was a center for a variety of industries, particularly the manufacture of high-quality chariots fashioned from wood and bronze, which revolutionized warfare early in the second millennium. What the chariots did, of course, was vastly improve the mobility of armies.

Twenty percent of the urban population, which was perhaps as large as 100,000, were artisans, including carpenters, gem cutters, and smiths, all of whom received their pay in the form of food, clothing, and oil. It is recorded that perfume workers, for example, were paid 2½ pints of oil daily, which they used to exchange for edible fats and soap. Copper and bronze workers in

Mari turned out swords, plows, pots, pans, and even tweezers. What seems most incredible is that we have such detailed information about a time when the historical part of the Bible is just beginning, with Abraham, the patriarch of the Jewish faith, leaving the southern Mesopotamian city of Ur and embarking on his journey to the land of Canaan.

In Mesopotamia during the third and second millennia B.C.E., cities, or more accurately city-states, agriculturally self-sufficient, were sophisticated trading centers boasting multitudes of artisans. A system of government evolved under kings who were believed to have descended from heaven. Codes of law developed. The prosperity and security of these societies came to be based to a large extent on tools and weapons crafted from a substance with unique and wonderful properties: metal.

2

The Age of Metals: A Primer

> And they shall beat their swords into plowshares, and
> their spears into pruning hooks.
>
> — Isaiah 2:4

ADAM, THE BIBLICAL FATHER OF HUMANITY, was
formed from earth, so it seems appropriate that the story
of the interconnections between materials and human
culture begin with stone and clay, substances taken directly from the earth. As human society became more
complex, so too did the materials it used.

Stone and clay's utility is severely limited by their
brittleness; they were poorly suited for tools that would
be subjected to bending, such as adzes. Metal, by contrast, could be easily shaped into thick or thin sections
by hammering (and in certain cases by casting) and used
in applications that required bending. For the first time,
craftspeople were free to devise instruments best suited
to the particular task at hand. As artisans learned how

Stephen L. Sass

to extract and work metals, a myriad of new or improved implements emerged from their workshops. One example was the plow, invented by the Sumerians. The first plows were likely made from tree branches; later they were fashioned from stone. But when copper was finally available and the resultant metal plow harnessed to oxen, the area of land that could be cultivated grew enormously. This led to the birth of the city.

Modern cities stand as monuments to the remarkable properties of a diverse collection of materials: concrete, glass, ceramics, and metals. And among metals steel is perhaps the most critical. Anyone who has been in Hong Kong or Beijing in the past decade has encountered the steel skeletons of half-finished buildings, towered over by huge T-shaped cranes outlined against the sky. Supported by their steel girders, these structures are testimony to the advantages of an endoskeleton over an exoskeleton. Early buildings, from the humble mud huts of Çatal Hüyük to the soaring dome of the Pantheon in Rome, carried their loads by means of an exoskeleton, like those of spiders and lobsters, with thick, load-bearing external walls. Late in the nineteenth century, architects came to appreciate that the internal skeleton of humans provided a more efficient model for bearing loads. This insight, and the discovery of new processes to manufacture large quantities of steel cheaply, made modern cities what they are.

New York's skyscrapers depend upon a latticework of girders to transmit the weight of the upper structure to the foundation. Without metals there would be no giant cranes to erect these buildings. Architects would have to resort to designing massive lower walls to bear the weight of the upper structure — the favorite construction technique of antiquity, best exemplified by the ziggurats of Sumer and the pyramids of Egypt. Our cities would be collections of low, squat buildings. And windows would be rare, because they weaken the walls. Medieval buttresses could reinforce tall load-bearing walls and help them achieve height, but they would present practical problems in the crowded downtowns of our modern cities. And were iron beams used rather than steel, girders would need to be three to five times greater in cross-sectional area and weight. Without steel, the architecture of our bridges would be quite different: rather than magnificent gossamer spans — like those of that icon of nineteenth-century American ambition, the Brooklyn Bridge — massive pillars would bear the crossing road.

Yet today steel is frequently supplanted by aluminum. Aluminum alloys compete with steel if weight is a factor. To make automobiles lighter and therefore more energy efficient, aluminum engine blocks are replacing cast iron. And, of course, high-strength aluminum alloys have been the material of choice for aircraft fuselages for more than fifty years. Tennis rackets perhaps best illustrate the rapidity with which one material displaces another. Twenty years ago steel replaced wood, and ten years ago graphite fiber–reinforced composites displaced steel. Since aluminum is light and strong, one might wonder why it did not replace steel in tennis rackets. Strength is not the only criterion when designing load-bearing structures. A tennis racket also needs to be reasonably stiff. Aluminum is not as stiff as steel, since it bends more than twice as much as steel does under the same load.

But I am getting ahead of myself, bandying about terms like "strength" and "stiffness" without explaining what they really mean. To appreciate the crucial role played by metals such as iron and bronze in early civilizations of the Near East, and to understand the reason they lent their names to eras, we have to look at their properties.

Metals are strikingly different from stone and ceramic in many ways. They conduct electricity and heat much better than these earlier substances. Good electrical conductivity was of no importance, however, until the nineteenth century. The primary reason metals displaced earlier materials was because of their responses to applied loads. Generally speaking there are three possible responses: *elastic deformation,* a change in length with load that is completely reversible upon unloading, such as when you stretch a rubber band and then let go (rubber bands undergo large elastic deformation; metals do not); *plastic deformation,* a change in length or shape with load that remains upon removal of the load, such as when you bend a paper clip; and *fracture,* breaking into two or more pieces under load, such as when you drop a glass.

The degree of "stiffness" is a measure of the load required to produce a particular elastic deformation and is calculated by dividing that load by the magnitude of the deformation it creates. Doing so, you get what we materials scientists call the *Young's modulus* of elasticity, the ratio of the stress applied to a body to the strain it produces (named for Thomas Young, 1773–1829, a British scientist, physician, and Egyptologist who was also the first person to show that light has a wave nature). The area bearing the load affects the response of the whole solid, which will be quite different for specimens that are one inch by one inch as opposed to those one foot by one foot. To account for this difference, the load is divided by the area, a calculation that gives "stress." The greater the stress needed to produce elastic deformation, the greater the stiffness. So when steel is said to be stiffer than aluminum, this means that it has a higher Young's modulus. Stiffness is a critical property when building a bridge, for example, which should not elongate or deflect an appreciable amount when a heavy truck passes across it. Imagine a bridge roadway that sank two feet as you drove across it and then sprang back to its original shape. Young's moduli for metals are typically more than one thousand times larger than for rubber (one of many reasons why metals and not rubber are used to build bridges). A prime design criterion for a load-bearing structure is that the amount it deflects under a load be minuscule. You cannot eliminate displacements entirely, and even large structures, such as the Empire State Building, are designed to flex elastically in strong winds.

A metal's strength is a measure of the load it can bear before permanently changing shape, or yielding. Strength is often referred to as "yield stress," a strong metal having a high yield stress and a weak metal a low yield stress. When we say that steel is stronger than iron we really mean that its yield stress is higher and that it resists plastic deformation better. When engineers design a load-bearing structure, the yield stress is of crucial importance. A bridge that lengthens (or sags) permanently whenever a heavy truck drives across it isn't very practical. Similarly, an airplane wing that flexed upward during flight and locked in that position would be an aeronautical engineer's nightmare. Therefore, another crucial design criterion for any metallic component is that the load it is supposed to bear must never exceed its yield stress.

Metals such as copper have the remarkable ability to change shape permanently during hammering or bending without fracturing or crumbling to pieces. If you tried hammering or bending ceramics, they would shatter. Copper's malleability meant that smiths could in fact "beat swords into plowshares" or sickles. To shape a ceramic sickle, on the other hand, meant first modeling the clay, then carefully drying it at low temperatures, and finally firing it at high temperatures. In addition to being easier to fabricate, metal sickles were also easier to sharpen and had much better fracture resistance. Drop a clay sickle on the floor and it will shatter. The site at Tel al-Ubaid, a late-fifth millennium B.C.E village not far from Ur, is littered with fragments of clay sickles. We can take small comfort from Mesopotamia's having been a "disposable society" six thousand years before our own.

I've already mentioned that pure metals are quite weak. But they can be strengthened by plastic deformation in a process called "work-hardening." Early smiths took advantage of this unique phenomenon to harden their copper weapons and tools by hammering them.

Without this strengthening process, early metal implements would have been useless, losing their edge the first time they ran into anything hard.

As to the reasons a metal gets stronger when it is hammered, research over the past seventy years has taught us a great deal about work-hardening processes, particularly as they involve the underlying atomic structure. The atomic arrangement in crystals is described by a basic building block called a "unit cell." In the case of copper the unit cell is cube-shaped and has atoms at its corners and at the centers of each face, and hence is termed "face-centered cubic" (or FCC). This basic building block is repeated at regular intervals — periodically — in three dimensions, taking the shape of a crystal. The beautifully symmetrical cube- and octahedral-shaped minerals found in rock shops are built up by this periodic process. The length of the edge of the unit cell for copper is 3.6 angstroms (named for the Swedish physicist Anders Ångström), or 0.000000036 centimeters, since one angstrom (abbreviated Å) equals 0.00000001 centimeters, which is one ten-billionth of a meter, so that a cube-shaped copper crystal with sides one centimeter long has twenty-five million unit cells along each edge.[3]

Such a crystal, if perfect, has a very high strength. Theoretically it should be able to withstand a stress of 500,000 pounds per square inch without permanently deforming. Moderately pure copper, however, has a yield stress far below its theoretical strength and can withstand only between 1,000 and 10,000 pounds per square inch before it permanently changes shape. If copper is purer, for example 99.999 percent pure (meaning that one atom out of every 100,000 is not copper), this makes the situation even worse: copper's yield stress can fall as low as a few hundred pounds per square inch, a far cry from the theoretical value. This enormous discrepancy between the theoretical and experimental strengths of copper was disconcerting to scientists until the 1930s, when they speculated that crystals are not perfect, but full of defects called "dislocations."

To visualize a dislocation, imagine a particular plane of atoms. Instead of going all the way from one side of the crystal to the other, this plane ends abruptly within the crystal, its edge running in a line down the whole length of the crystal. The gap weakens the crystal. Dislocations in metals move very easily under an applied load, allowing the crystals to change shape, even under very low stresses. When a dislocation moves from the left-hand side of the crystal to the right-hand side, the extra plane of atoms emerges at the surface, creating a step; when many steps form on the surface of the crystal, it changes shape permanently. The strength of crystalline metals depends upon dislocations. The ease with which these dislocations move will determine the stress at which plastic deformation takes place. Elastic deformation, involving the stretching or compressing of bonds between atoms, is reversible, while plastic deformation is not.

Materials scientists are able to produce only tiny whiskers of nearly perfect materials, such as iron, so small they have diameters of 0.0001 cm, but which do possess the theoretical strength I mentioned earlier. For a long time these whiskers were considered laboratory curiosities. An object with a diameter one-fiftieth of that of human hair would seem to have few practical applications. But recently, large-diameter whiskers have been suspended in a matrix of another material, making what we call a "composite material," something I will discuss at a later point.

Crystals containing dislocations can be strengthened by putting obstacles in the way of their motion. Just as it takes an extra effort for runners to clear the gates in the 100-yard hurdles, it takes extra stress for dislocations to overcome obstacles. One such barrier is another dislocation. Dislocations can be deliberately formed during plastic deformation. For a dislocation moving on a particular atomic plane to traverse the crystal, it must cut through a tangle of dislocations. This takes more stress than if the crystal were nearly dislocation-free. Hence, additional dislocations make the crystal stronger. Imagine a car moving along a road with no traffic, then suddenly coming to a gridlocked intersection. During plastic deformation not only do the dislocations move, they also multiply, sometimes as much as a thousand-fold. They become even more tangled, and the more tangled the stronger the crystal.

This is the physical basis of the process of work-hardening, in which a network of dislocations is literally hammered into the material.

What makes metals unique among all substances is that when they are heated to temperatures well below their melting point (or "annealed"), the strength given them by addition of dislocations can be wholly removed, *without* affecting the change in shape that deforming them has produced. A suitable annealing temperature for copper is 600 degrees, easily obtainable in the furnaces of the ancient world several thousand years before the time of Christ. When that temperature is reached, the crystal's dislocations move around and destroy one another whenever they collide, forming continuous planes and a locally perfect crystal. The plastic deformation and annealing cycle of metals is of enormous importance, since heavily deformed metals, while very strong and therefore resistant to further permanent deformation, are also quite brittle. The same processes that give metals high strength also make them more susceptible to breaking — nature's way of saying that you can't have your cake and eat it too. To roll a brick-shaped ingot down from two inches to 0.02 inches in thickness — as one might do to make a car body, for example — requires lots of plastic deformation and plenty of annealing cycles, so that the process can take place without fracture. Hard, brittle copper can be made soft and malleable again and again by heating it for an hour at 600 degrees after plastic deformation. Ceramics cannot be made soft again once they are fired. Metals can.

Most engineering materials we deal with do not consist of one single crystal but many crystals, making up together a polycrystalline solid. The plane on which two differently oriented crystals meet is called a "grain boundary." Many grain boundaries contain dislocations, so they also act as barriers to the motion of other dislocations. In fact, another way to strengthen pure polycrystalline metals is to make the average size of the individual crystals very small. This increases the number (and area) of grain boundaries, and therefore the number of barriers to dislocation motion. Grain size control is widely used today to tailor mechanical properties, but it was unknown to the smiths of the ancient Near East.

Other barriers can be introduced to hinder the motion of dislocations. Small hard particles called "precipitates" are also excellent strengthening agents. Their presence gives rise to "precipitation hardening," the strengthening mechanism in the aluminum alloys for aircraft fuselages and the nickel superalloys used to make jet engine turbine blades (all to be discussed later).

Now that I've outlined the unique properties of metals and their origin, I'll turn to how humans first came to recognize and work them. The first metals to have a strong impact on human existence were copper and bronze, and they are the place to begin.

3

Copper and Bronze

> And there came out from the camp of the Philistines
> a champion named Goliath, of Gath, whose height
> was six cubits and a span [ten feet]. He had a helmet
> of bronze on his head, and he was armed with a coat
> of mail, and the weight of the coat was five thou-
> sand shekels of bronze [ninety pounds]. And he had
> greaves of bronze upon his legs, and a javelin of
> bronze slung between his shoulders.
>
> — 1 Samuel 17:4–7

ANCIENT PEOPLES EXPLOITED what they could find
around them: wood, stone, and clay. Only within the
past eight to nine thousand years have humans recognized and begun experimenting with metals. We might
wonder why it took so long. Anyone who hikes knows
that wood, stone, and clay can be found scattered across
the countryside, there for the taking. Not true for metals. The following tabulation of the abundance of metals

in the earth's crust (including two additional non-metal chemical elements, oxygen and silicon, for comparison) illustrates how much of a challenge finding certain metals is:

Metal	Abundance in the Earth's Crust (Weight-percent)
(Oxygen)	46.0
(Silicon)	28.0
Aluminum	8.0
Iron	5.8
Magnesium	4.0
Calcium	2.4
Potassium	2.3
Sodium	2.1
Copper	0.0058
Lead	0.0001
Tin	0.00015
Silver	0.000008
Platinum	0.000004
Gold	0.0000004

To exploit copper successfully, early artisans needed to find it in highly concentrated forms — either nearly pure or in ores (minerals made up of several chemical compounds) with high metallic content, typically containing copper oxide, copper carbonate, and copper sulfide. Rich deposits were important even for iron, where a valuable ore might contain 50 percent metal. Because iron makes up approximately 5 percent of the earth's crust, our ancestors had to find it in ores enriched by ten times if they were going to extract it economically. Rarer metals like gold and silver needed to be concentrated more than a thousand times to be recoverable.

Because the earth's crust contains so much more aluminum and iron than copper, we might wonder why copper was the first metal to revolutionize our use of materials, nearly three thousand years before iron and more than five thousand years before aluminum. Abundance clearly played no role. Copper's early exploitation can be explained by the ease with which it could be obtained or smelted from ores. Commercial aluminum and iron ores are what are termed "oxides,"compounds containing metal and oxygen atoms. Aluminum oxide is more stable than iron oxide, meaning that it is much more difficult to break down into its component elements and therefore that aluminum is more difficult to extract from its ores than iron. That stability is the same reason aluminum is more resistant to corrosion than iron. When bare aluminum is exposed to oxygen in the air, it rapidly forms a thin but tenacious oxide film that protects the underlying metal from further attack by the atmosphere. Iron ores in turn are more stable than copper ores, which are the easiest to "reduce," or smelt, back to metal. Smelting takes advantage of the class of chemical reactions called "oxidation-reduction." For example, when one molecule of the solid copper oxide, Cu_2O, reacts with one molecule of gaseous carbon monoxide, CO, it forms two atoms of copper, 2Cu, and one molecule of gaseous carbon dioxide, CO_2. This is a "reduction" reaction, with copper oxide reduced back to pure copper, as a result of carbon monoxide's strong desire to acquire one more oxygen atom to form carbon dioxide. The reverse of this is an "oxidation" reaction, during which copper reacts with oxygen to form copper oxide.

Copper was used extensively before iron because artisans found ways to extract it from its ores first. As early as the seventh millennium B.C.E., they were crafting copper into trinkets and then into more substantial objects, including tools and ceremonial articles. Metal workers fashioned them from relatively pure copper, found in greenish nodules scattered in mountain streambeds. This pure or "native" copper generally occurs in small quantities, though occasionally larger amounts have been discovered. In the 1700s, a fur trader stumbled upon a three-thousand-pound copper boulder on the southern shore of Lake Superior in Michigan. Today it resides in the Smithsonian Institution in Washington, D.C.

So long as artisans were forced to rely on native copper, they were limited to fashioning small pieces of jewelry and cult articles. There were, however, copper-rich minerals in the mountains north of Mesopotamia, stretching from Anatolia to the Caspian Sea; at Timna in the southern part of the Negev Desert, below the Dead Sea; and on Cyprus in the Mediterranean. The word *copper* comes from the Latin word *cuprum,* or "Cypriot metal," a reference to the large ore deposits found on the island. To obtain sufficient quantities of copper for tools and weapons, early craftspeople had to make two crucial discoveries. First they had to learn that brittle, rocklike minerals contained shiny, malleable copper, and second they had to unravel the secrets of smelting.

Metal workers first smelted copper just after 4000 B.C.E., in what today is Iran. We can only speculate how they extracted their precious prize from its ores. Copper-rich minerals, such as dark-green malachite, are often beautiful, and so early craftspeople probably sought them for use in jewelry. Malachite contains copper, carbon, oxygen, and hydrogen. By heating it with charcoal — an excellent reducing agent because the carbon reacts strongly with oxygen in the ore — metal workers were able to drive off gaseous carbon dioxide and water, leaving behind metallic copper. Cooking fires were not hot enough to smelt copper, so the early reduction of ores most likely took place in pottery kilns, which could reach the desired temperature of 1200 degrees. The first smelting of copper might have occurred when a potter accidentally dropped malachite into his kiln. The malachite then reacted with the partially burned wood and was reduced to copper. Imagine the excitement of an astonished potter as he digs out chunks of pure copper while cleaning his kiln — during a time when the only copper available was found scattered in streambeds and was used primarily for small, extremely valuable trinkets. A variation on this story might involve pieces of malachite being pressed as decorations into clay pots, which were then fired. Transforming a greenish stone into copper must have seemed like magic. What power rested in the hands of those who could control this process!

Whether or not copper was first smelted in a kiln, serendipitous discoveries inspired experimentation, as they do in all ages. Once smelting was discovered, vast new worlds beckoned artisans to explore. Ancient craftspeople tried to convert all manner of minerals into valuable metals. The floodgates were opened. If copper, why not gold? Alchemy, based on the belief that base substances could be turned into gold and silver, doubtless emerged from these early successes at smelting. Modern chemistry, and all of the wonderful materials that have come from it, traces its roots back through alchemy to the smelting of ores. We owe much to our hypothetical potter.

While smelting gave us metals, it also created a host of practical problems. Smelters required at least 220 pounds of charcoal to extract eleven pounds of copper, which would be needed to fashion ten to twenty ax heads. Wood could be converted into charcoal by heating it in covered pits, driving off water and volatile resins, and leaving behind a porous mixture of carbon and ash. Each pound of charcoal required seven pounds of wood, so to produce one pound of copper took 140 pounds of wood. As the demand for metals grew, smelters became voracious consumers of wood. Several millennia later, their insatiable appetite contributed to a timber shortage in England. As their forests rapidly disappeared, the English fortuitously discovered a new source of carbon in the form of coal, which fueled the Industrial Revolution.

In ancient times carbon in the form of charcoal was the key to a metals-based society, but today it is relegated to backyard barbecues, artists' studios, and fishtank filters. However, carbon as graphite, diamond (industrial, not gem), and coke still plays a critical role. Carbon also powers our world in the form of coal and oil, and sup-

ports our material culture with polymers and graphite fiber–reinforced composites. We are all largely made
up of carbon-based polymer molecules. Nature is bounti-
ful in wood, coal, and oil, the sources of materials and
energy. Just as the peoples of the seventeenth and
eighteenth centuries turned increasingly to coal as a substitute for wood, so it will soon be time for us to look for
a replacement for oil, the reserves of which will run out
sometime in the next century.

History teaches us that the depletion of one source
of a substance leads to the discovery of a new source, or
its replacement by an entirely new material. As the demand for copper implements grew in the Near East, deposits of oxide and carbonate ores became exhausted,
and smelters had to turn to sulfur-rich ores. These were
more difficult to smelt, requiring an additional "roast-
ing" step to burn off most of the sulfur as sulfur dioxide,
leaving behind a mixture of oxides and sulfides suitable
for final reduction to copper. While a complicated chore,
smelting sulfide ores would later come to play a crucial
role in fostering the birth of the Iron Age at the close of
the second millennium B.C.E., because these copper-rich
ores, fortuitously enough, also contained iron.

In their search for valuable minerals, ancient prospectors used the most advanced technology of the day,
such as a forked willow dowsing or divining rod. They
also depended on their senses, smell in particular. Gar-
licky odors identified arsenic-rich ores when they were
struck by a pick; copper ores frequently contain arsenic.
Prospectors also knew that a dwarfed form of juniper
often grew in regions containing copper ores. The first

scientific treatise on mining and metallurgy, *De Re Metallica,* was written in 1556 by Georgius Agricola, a German mining expert, scholar, and physician. It was later translated by Herbert Hoover, a mining engineer and future president of the United States, with the help of his wife. Agricola describes how prospectors identified various minerals:

> The waters of springs taste according to the juice they contain, and they differ greatly in that respect. There are six kinds of these tastes which the worker usually observes and examines: there is the salty kind, which shows that salt may be obtained by evaporation; the nitrous, which indicates soda; the aluminous kind, which indicates alum; the vitroline, which indicates vitriol; the sulfurous kind, which indicates sulfur; and as for the bituminous juice, out of which bitumen is melted down, the color itself proclaims it to the worker who is evaporating it. . . . Therefore an industrious and diligent man observes and makes use of these things and thus contributes something to the common welfare.[4]

Returning to ancient times, copper and turquoise for the pharaohs of Egypt were mined in the southern part of the Sinai Desert. The quest for such prized substances led to the Egyptians' early conquests and the formation of their empire. During the late nineteenth century B.C.E., scribes recorded how these mines tested the endurance of both administrators and diggers. Inscriptions by Hor-ur-Re, a high government official of Pharaoh Amenemhet the Third of the Twelfth Dynasty, found at Serabit el-Khadim in the Sinai, vividly illustrate the brutal conditions suffered by Egyptian miners, who toiled under the searing sun before unguents and potions were available to protect them:

> The Pharaoh dispatched the Seal-Bearer of the God, the Overseer of the Cabinet, and Director of Lances, Hor-ur-Re, to this mining area. This land was reached in the 3rd month of the second season [June], although it was not at all the season for coming to this mining area. The Seal-Bearer of the God says to the officials who may come to this mining area at this season:
> Let not your faces flag because of it. Behold ye, Hat-Hor [Egyptian goddess of the Sinai mines] turns it to good. I have seen it so with regard to myself. I came from Egypt with my face flagging. It was difficult, in my experience, to find the proper skin for it, when the land was burning hot, the highland was in summer, and the mountains branded an already blistered skin. When the day broke for my leading to the camp, I kept on addressing the craftsmen about it: "How fortunate is he who is in this mining area!" But they said: "Turquoise is always in the mountain, but it is the proper skin which has to be sought at this season. We used to hear the like, that ore is forthcoming at this season, but, really, it is the skin that is lacking for it in this difficult season of summer!"[5]

While excavating a cave in the hills above the Dead Sea at Nahal Mishmar, archaeologists discovered artifacts revealing some of the earliest uses of copper. Bedouins had first stumbled on the caves, drawing the attention of Israeli archaeologists, who systematically explored levels beneath the cave floor dating from the second century C.E. (the time of the Bar Kokhba Revolt of the Jews against the Romans) back to the fourth millennium B.C.E. (the Chalcolithic Period — *chalkos* being Greek for "copper"). They unearthed a cache today called the Judaean Desert Treasure, including a spectacular collection of primarily metal objects from the Chalcolithic Period. The uses and origins of these copper and copper-arsenic alloy artifacts are still open to speculation. Perhaps they were looted from a shrine, or perhaps they belonged to a temple at nearby En-gedi and were hurriedly buried to escape being looted by a pharaoh.

Many of the artifacts were fabricated by the "lost wax" technique, in which a remarkably complex object can be cast in one piece. Other methods of casting include using an open mold or a closed mold, but the lost wax technique permitted detail and intricacy never before seen. An artisan would first craft a beeswax replica of a crown, for example, around a solid clay core, sculpting the surface of the wax. Then, packing clay around the wax to fill in all the details of the model, he would carefully leave holes at the top and bottom. After the clay dried, it was buried in sand and fired. During firing the wax either burned, or melted and ran out of the holes — hence "lost" — leaving a ceramic mold ready for casting. After pouring in liquid metal, which cooled and solidified, the copper worker then broke open the mold. The "crown" found in the Judaean Desert is a particularly beautiful example of the high level of technical expertise of artisans of the fourth millennium B.C.E. The ancient lost wax technique is still used today and has important high-technology applications, including the production of single-crystal blades for jet turbine engines.

The treasures from the Judaean Desert represent the acme of the metallurgical skill of Chalcolithic artisans, working from 4500 to 3150 B.C.E., when copper and copper-arsenic alloys were first widely used in the Near East. Adding tin to copper to form bronze began in the thirty-second century B.C.E., the advent of what we call the Bronze Age, but copper and copper-arsenic alloys remained the dominant metals for tools and weapons for another thousand years. Naming the ages of human history for materials is clearly more an art than a science.

Bronze and copper-arsenic alloys are typically 90 to 95 percent copper by weight, with a small amount of a second component, tin. Why did smelters begin adding tin to copper? Dissolving one metal in another is much like dissolving sugar in water. Coppersmiths could make the alloy by first dissolving the tin into molten copper, and then letting it cool and crystallize. But why add any second component, and especially tin, which was a sacred commodity in the ancient Near East?

First, a technical explanation. We know pure metals are intrinsically weak, with yield stresses of less than 1,000 pounds per square inch. In antiquity, forming an alloy had two benefits: it strengthened the metal by a process called "solid solution hardening," and it made casting easier by lowering the melting temperature. Solid solution hardening occurs when the dissolving (or solute) atom is either larger or smaller than the host (or solvent) atom. For example, copper and arsenic atoms have diameters of 2.56 Å and 2.51 Å, respectively. Because arsenic is smaller than copper, it is attracted to the dislocation line, at which the extra plane of atoms is squeezed by the two adjacent planes and the atomic spacing is smaller than elsewhere in the material. In other words, the smaller arsenic atoms relieve the local crowding. When such a dislocation is put under an applied load, the cluster of arsenic atoms at the end of the extra plane prevents it from moving, making the metal stronger.

Native copper was rare, particularly in the vast alluvial Mesopotamian plain, which is why the transition from clay and stone to metallic implements occurred so slowly. It took several thousand years after copper was first found before there were sufficient amounts of it to craft into something useful. Early merchants did a thriving business in copper, which they traded in the form of sheets shaped like an ox hide (recognizable for trading purposes) typically weighing sixty to seventy pounds. By the early third millennium B.C.E., Mesopotamian artisans were fashioning vases, mirrors, daggers, hoes, and plows from imported copper. Metal implements were so valuable that they are rarely found in digs, unlike ceramic artifacts, which archaeologists find in abundance. This is because copper was constantly being recycled. A cuneiform record from Ur informs us that smiths had received "1,083 copper sickles and 60 copper hoes" from the storehouse for refurbishment.[6]

The earliest copper artifacts contained small quantities of arsenic, not because the smiths added it, but because it was an impurity in common copper ores. Artisans could strengthen copper tools and weapons by both solid solution and work-hardening. Bronze is stronger than pure copper and, for a ten-weight-percent tin composition (in other words, ten grams of tin added to ninety grams of copper), also easier to cast, since its melting point is lower than that of pure copper by 80 degrees. (The lowering of the freezing point of a liquid by the addition of a second component is common practice. Salting roads melts ice by lowering its freezing temperature below zero.) Moving from copper-arsenic to copper-tin alloys gave little improvement in mechanical properties, however, since both could be work-hardened for tools and weapons, though bronze implements did have better properties after casting.

The problem with arsenic was that it was lost from copper ores during smelting, because heating with air forms arsenic trioxide, a volatile compound that is easily sublimated, meaning that it goes directly from solid to gas. Arsenic's volatility led to wide variability in alloy composition and, consequently, in mechanical properties. In contrast, when smiths formulated bronze by combining copper with tin from cassiterite — a relatively pure tin oxide ore — they were confident that its composition and properties were reproducible from batch to batch.

Copper-arsenic ores, such as enargite (a gray-black

sulfide), were relatively scarce. Artisans probably stumbled upon bronze after they had run out of copper-arsenic ores and were forced to smelt arsenic-free ores. While plentiful, these ores yielded copper with poor properties. In their search for other arsenic-containing ores, prospectors may have mistaken stannite, a copper sulfide compound also containing tin and iron, for enargite; stannite gave alloys properties similar to those of copper-arsenic. But the combination was in fact bronze.

Early alloys often have a number of "origin stories," since conditions inevitably varied from region to region, and Egyptian tomb paintings dating from the end of the second millennium B.C.E. tell a different tale about the first use of bronze. They suggest that metal workers intentionally melted ingots of copper and tin together. Cuneiform texts from Mari also support this version: "One-third mina of tin to 2⅔ minas of washed copper from Tema has been alloyed at the ratio of eight to one. Total: three minas, ten shekels of bronze for a key."[7] This tells us explicitly that tin was added to copper to form bronze. Whatever their reasons for putting copper and tin together, and however they discovered it, once artisans knew about bronze, it was their preferred material for fabricating tools and weapons for more than a thousand years.

Working with copper exacted a high price. Early copper workers put their lives at risk by breathing poisonous arsenic trioxide fumes venting from their furnaces. As arsenic accumulates in the body, it damages nerves and causes muscles to atrophy. Coppersmiths and smelters were usually crippled and led short lives. In his description of the Greek gods in the *Iliad,* Homer portrays the metalworker Hephaestus as lame. A symptom of arsenic poisoning, almost certainly. Vulcan, his Roman counterpart, was also crippled. Since tin is physiologically harmless, the displacement of copper-arsenic alloys by bronze was a health benefit for copper workers.

Whatever risks they ran, bronze workers in antiquity developed remarkable expertise in the casting of monumental sculptures. Working with Solomon to erect his temple in Jerusalem during the tenth century B.C.E., Hiram of Tyre cast bronze pillars twenty-three cubits (thirty-eight feet, one cubit being about twenty inches) high and a bronze basin that was, as the Book of Kings describes it:

> Ten cubits from brim to brim, and five cubits high, and a line of thirty cubits measured its circumference. It stood upon twelve oxens, three facing north, three facing west, three facing south, and three facing east. . . . In the plain of Jordan the king cast them, in the clay ground between Succoth and Zarethan. And Solomon left all the vessels unweighed, because there were so many of them; the weight of bronze was not found out.

Two and a half centuries later, the Assyrian king Sennacherib, who reigned from 704 to 681 B.C.E., modestly described his "Palace Without Rival," situated on the crest of a central hill in Nineveh:

Eight lions, open at the knee [free standing], in the
posture of advance, which were cast of 11,400 talents [750,000 pounds] of bright copper. With two
colossal pillars whose copperwork came to 6,000
talents [400,000 pounds]. . . . Four mountain sheep,
as protecting deities, of silver and copper. . . . Following the advice of my head and the prompting of
my heart, I fashioned a work of bronze and cun-
ningly wrought it. Over great posts and crossbars of
wood, twelve fierce lion-colossi together with twelve
mighty bull-colossi, complete in form, twenty-two
cow-colossi clothed with exuberant strength and
with abundance and splendor heaped upon them, at
the command of the god, I built a form of clay and
poured bronze in it, as in making half-shekel pieces,
and finished their construction.[8]

If we are to believe Sennacherib, each lion weighed
forty-seven tons and was nearly three times lifesize,
while each copper pillar weighed 100 tons, slightly less
than a fully loaded Boeing 757 at takeoff. Stone carvings
from Ashurbanipal's palace at Nineveh, done fifty years
later, depict Sennacherib's palace exactly as he described
it, with bronze columns set on the backs of lions.

In 587 B.C.E., four centuries after Solomon erected
his temple, the Babylonian king Nebuchadnezzar the
Second pillaged Jerusalem. The tale recounted in 2 Kings
includes descriptions of the same marvelous metalwork:

And the pillars of bronze that were in the house of
the Lord, and the stands and the bronze sea that
were in the house of the Lord, the Chaldeans [Baby-

lonians] broke in pieces, and carried the bronze to Babylon. . . . As for the two pillars, the one sea, and the stands, which Solomon had made for the house of the Lord, the bronze of all these vessels was beyond weight. The height of the one pillar was eighteen cubits [thirty feet], and upon it was a capital of bronze; the height of the capital was three cubits [five feet].

Following the fall of Jerusalem, the Israelites were sent into their Babylonian exile, and their extraordinary bronzes looted and melted down in the furnaces of their oppressors.

Monumental castings of the sort described in the Bible in Kings and by Sennacherib, while certainly impressive and flattering to the egos of rulers, were built on the backs of laborers, literally and figuratively. Materials have eras named for them not because of their role in glorifying kings, but because they had significant impacts on the everyday life of common people, and by extension, on their culture. Gigantic castings were little more than gigantic banes to the existence of common folk. How did copper and bronze improve their lot?

When artisans fashioned arrowheads, sickles, hammers, daggers, and axes from stone, clay, or bone, they did so one at a time. Metalworkers, however, could turn out tools, farm implements, and weapons far more rapidly, and with superior properties, by first casting the objects to the desired shape and then hammering their edges to make them hard and sharp. This was another early attempt at mass production. And bronzesmiths could easily straighten and sharpen metal tools if they became bent or dull. Most significantly, tools were now given forms that best fit the task at hand. For example, carpenters prefer adzes with very thin blades, allowing them to remove thin strips of wood for their precise carvings. Such a tool is easily formed from metal. A stone blade of small dimensions gave a far less precise adze, which easily broke or chipped. Metal plows were also superior to those made from wood and stone, increasing the amount of land that could be cultivated, and thus feeding the large populations of the early Mesopotamian cities of Ur, Uruk, and Eridu, as well as producing surpluses for trade.

Not just plows improved. Farmers and drovers doubtless had difficulty controlling the horses, asses, and oxen used for pulling plows and carts because leather bits could not survive the grinding of teeth and saliva. Bronze bits or cheek pieces were an excellent solution to this problem. Most important, craftspeople could fashion totally new tools; for example, some that were nearly impossible to fabricate from earlier substances, such as saw blades and drills.

And then there was war. Bronze made metal armor practical, although it was not always successful as the Old Testament tells us in the account of David and Goliath, from the early first millennium B.C.E. Rejecting Saul's offer of armor because it was too heavy, David put his faith in a staff, a sling, and five smooth stones, harkening back to earlier times. Goliath's massive bronze armor was no defense against a well-slung stone to the forehead. But, favored by the divine, David is a special case, and in most conflicts of the era bronze won the day.

Copper and its alloys dominated the ancient world the way steel and aluminum do today, but other metals had also been identified and exploited. I'll turn now to two that are inexorably linked to human greed.

4

Gold, Silver, and the Rise of Empires

"O accurst craving for gold!"

— Virgil, *Aeneid*

OUR ANCESTORS KILLED, and even destroyed entire civilizations for gold and silver — not because these metals were of any real use to them, but for their rarity and beauty. Iron and copper tarnish by oxidizing. In fact, for a long time iron was considered a black metal because it reacts rapidly with air to form blackish iron oxide or rust (hence the term "blacksmith"). Gold, by contrast, does not react with oxygen, and its yellowish-orange color and luster give it special allure. Gold is one of the noble metals, so called because of their resistance to corrosion, and we might readily agree, given its appearance and value. Being malleable, gold is easy to shape, allowing smiths to hammer it into sheets as thin as 0.0025 millimeters. Excavations of the royal tombs at Ur, dating from 2600 B.C.E., reveal both the wealth of the aristocratic class that emerged to rival the temple priests of Sumer and the remarkable skill of their goldsmiths. Among the artifacts are a helmet showing the intricate details of contemporary Sumerian hairstyles, and a lapis lazuli–adorned dagger, so modern-looking that archaeologists believed at first it was the handiwork of a thirteenth-century Arabian workshop.

Ancient prospectors found gold either in veins of quartz or as nuggets freed by erosion. The rounded appearance of nuggets comes from the tumbling action of the stream on the malleable gold. Early miners devised a particularly novel method for extracting gold: first crushing the quartz, then suspending it in water, and finally passing the "slurry" over fleeces of sheep. The heavy gold particles settled and adhered to the grease in the wool. In the well-known Greek legend, Jason and a collection of Greek heroes called the Argonauts, named after the ship they sailed in, the *Argo*, sought to recover a Golden Fleece. There was a practical basis for that association.

On the whole, little gold was present in the Near East, except in Egypt, where miners worked more than one hundred sites in the Nubian desert ("Nubia" meant "land of gold" in Egyptian). The pure-gold casket of Tutankhamen, weighing a staggering 240 pounds, amply illustrates the vast quantities of gold commanded by the Egyptian pharaohs. The tombs of the pharaohs were pillaged over the centuries because of rumors that they were filled with gold. What the pharaohs most coveted led to what they most feared — desecration of their final resting places.

The story of the legendary Phrygian king Midas is an object lesson about humanity's greed for gold. Midas begged the gods to give him a golden touch, only to discover to his horror that it would lead to starvation. He then begged the gods that the gift be taken back: Midas had learned that there could be too much of a good thing.

Unlike gold, metallic silver rarely occurs in nature, since it reacts rapidly with gases in water and air to form silver sulfide compounds (or tarnish). Gold and silver are frequently found together in an alloy called "electrum," a favorite of ancient artisans for use in decoration, exemplified by a beautifully sculpted donkey, part of an ornamental horse bit unearthed from the royal cemetery of Ur.

By far the largest sources of silver in the ancient world were lead ores, but until metallurgists discovered how to smelt them very little silver was available to artisans. To extract their precious prize, smelters first reduced the ore to lead containing small quantities of silver. Then, in a process called "cupellation," they oxidized the lead in a shallow porous cup (a "cupel"), by heating it to temperatures above the melting point of silver, 961 degrees. The lead oxide vaporized and was absorbed in porous bone or ground-up pottery shards in the cupel, leaving behind a shining globule of silver. It would take three thousand years, however — from the seventh millennium B.C.E., when lead was first identified, to the fourth — before silver was finally won from lead ores. And even then low-grade ores and primitive technology meant miners were capable of extracting only one pound of silver from a ton of lead. At the beginning of the Common Era, the richest mines were capable of yielding merely ten pounds of silver from a ton of lead. We can only speculate how it was that silver workers in antiquity ever learned that lead ores contained tiny amounts of silver, and marvel at how they devised such an ingenious process as cupellation to extract these trace quantities. Greed, of course, is a powerful incentive for inventiveness. Unfortunately, the cupellation process is not only complicated but messy: much lead oxide vapor escaped from the cupels. Many of the lakes of Europe are polluted today with high concentrations of lead, thanks to the practices employed to smelt silver for the last three or four millennia.

Lead was one of the first metals to be smelted from its ore, typically a lead sulfide–bearing mineral called galena. Heating galena decomposes it to a mixture of lead sulfide and oxide, which then reacts at 800 degrees to form metallic lead. Lead is much weaker than copper and its alloys, and proved of little value for tools and weapons. In antiquity the primary uses of lead were as clamps to link blocks for building, sheets for roofing, and, alloyed with tin, as solder for joining together metal pieces. Late in the first millennium B.C.E., pipes of lead were introduced into the Greek and Roman water systems because it was so soft and easy to shape. Much speculation has focused on the role lead poisoning played in the decline of the Roman Empire. Water passing through these pipes picked up lead, which gradually accumulates in the nervous system, bones, liver, pancreas, teeth, and gums. Lead impairs red blood cell production, leading to anemia, with symptoms of fatigue, headaches, and dizziness. A high enough concentration in the blood can cause nerve damage and death.

Spurred by their early successes with lead, innovative metallurgists sought new ways to transform less valuable substances into gold and silver. If silver could be won from lead, why not from other base materials? Cuneiform inscriptions dating from the reign of Nebuchadnezzar I, king of Babylon during the twelfth century B.C.E. and distant ancestor of the king who drove the Israelites into exile, describe attempts at synthesizing silver by mixing copper, bronze, and an unknown mineral, as well as attempts to create valuable gemstones by heating powders made from common stones, red alkali, milk, mountain honey, and wine. These efforts were doomed to failure. Alchemists' dreams would only be realized in the early 1950s, when scientists succeeded in transforming graphite into diamond. Of course, this was not a matter of alchemy; rather, the scientists transformed one crystalline structure, or "phase" (a homogeneous material with a particular structure and composition) of carbon into another by applying what had been learned about the effect of temperature and pressure on the stability of the various phases of carbon.

Silver-rich galena deposits discovered at the end of the sixth century B.C.E. at Laurion, near Athens, played a decisive historical role by financing the rise of the city-state of Attica in the fifth century. Silver from these mines provided the Greeks with the means to build a large fleet of triremes, ships powered by three tiers of rowers, which were instrumental in their defeat of the Persian king Xerxes off the island of Salamis in 480. The destruction of the Persian fleet marked the turning point in the fortunes of the great Persian empire, which earlier had expanded so rapidly under Cyrus and Darius. In the mid-fourth century, Philip II's emergence in Macedonia presaged Persia's final collapse. Philip's access to the wealth of the gold mines at Mount Pangaeum in Thrace enabled him to create a nearly invincible professional army, employing a new infantry formation, the phalanx, which he used first to conquer and then to unify the Greek city-states. Following Philip's assassination in 336, his son, Alexander (who may well have had a hand in his father's death), went on to defeat the Persians in a campaign stretching from Egypt in the south to beyond the Indus River in India in the east. One important consequence of Alexander's conquests was the founding of Alexandria in Egypt, which eventually grew to nearly one million people and became the world center of learning and trade for centuries. It was here that scholars began translating the Old Testament from Aramaic and Hebrew into Greek.

Of course, not all these events can be directly attributed to the silver from Laurion and the gold from Pangaeum. Without these natural sources of wealth, however, the Greeks would likely have failed in their war with the Persians and the history of the eastern Mediterranean would have taken quite a different path.

The Greeks would have been even wealthier had they been able to extract silver from lead with greater efficiency. A third of the silver in the lead ore was left lying in waste dumps around Laurion. In the third century C.E. Roman smelters reworked two million tons of this slag, containing between 0.005 and 0.010 percent silver. It was reworked yet again in the 1850s.

Gold and silver also played a crucial role in the rise of Rome, heir to the remnants of Alexander's empire. The Romans established vast mining works at Rio Tinto in the Sierra Morena range in Iberia (Spain), applying their engineering expertise to smelting copper ores rich in gold and silver, which provided the economic foundation of their empire. By the reign of Caesar Augustus, Rome was a city of nearly a million people, yet had no means of feeding itself other than by importing 200,000 to 400,000 tons of grain annually from its colonies in Egypt and North Africa. Rome actually manufactured very little, and the wealth to pay for these basic foodstuffs came from taxes, military booty, and its mines.

Slaves condemned to work in Roman mines endured unimaginably miserable conditions. Some never saw the light of day for months on end. The second-century writer Apuleius, describing slaves working in a flour mill, provides us with a glimpse of their general degradation in his work *The Golden Ass:*

> Their skins were seamed all over with the marks of old floggings, as you could see through the holes in their ragged shirts that shaded rather than covered

their scarred backs; but some wore only loincloths. They had letters marked on their foreheads, and half-shaved heads and irons on their legs.

Continually pressured by their imperial masters to increase the output of the Spanish mines, Roman smelters pioneered a variety of innovative techniques, including "amalgamation," or dissolution of precious metals from pulverized ores into liquid mercury. Subsequent distillation of the mercury left behind gold and silver. The Romans obtained mercury from their mines at Almaden in Spain by heating a reddish-orange mineral called cinnabar, mercury sulfide, and decomposing it into sulfur dioxide and liquid metal. Much later, in the late 1700s, studies of the reaction of mercury with oxygen by the French scientist Antoine-Laurent Lavoisier would play a critical role in tearing away the veil of ignorance surrounding the elemental nature of matter and give a firm base to the discipline of chemistry.

Seeking to further improve the extraction of minuscule amounts of gold and silver from their Iberian mines, Roman smelters then devised the process called "liquation." Taking advantage of the greater solubility of gold and silver in lead than copper, smelters melted copper containing these metals, together with large quantities of lead, which absorbed them. Since copper and lead are not soluble in each other, and lead melts at a much lower temperature (328 degrees) than copper (1083 degrees), the lead and copper could be separated by gentle heating, allowing the liquid lead, laden with gold and silver, to drain off. Cupellation then recovered the precious metals from the lead.

Roman smelters mastered complex chemical processes to support their empire, and Rome's expansion, in turn, was driven by its insatiable appetite for new sources of valuable minerals. By the beginning of the second century, Rome controlled every known precious-metal mine within its far-flung empire.

Silver and gold played a crucial role in the mechanics of commerce long before the Greek and Roman empires. At the beginning of the Modern Era, in the ninth millennium B.C.E., trade essentially consisted of bartering one commodity for another. Problems arose when, for example, wheat traders did not need the lapis lazuli offered in exchange for their grain. Merchants could of course accept the precious stones in the hope of swapping them for something else of greater interest, but the perennial difficulty of matching the needs of importer and exporter hindered trade, particularly before there were written languages. As commerce expanded, so did a demand for a mutually acceptable and portable commodity whose value did not fluctuate — as wheat might, depending on the annual flooding of the Euphrates. Silver and gold soon became the favored mediums of exchange, because they kept their worth over long periods of time and were in relatively short supply. It was the beginning of the "gold standard."

Since silver could be stored, divided, and melted with ease into new shapes, it became the primary currency. Silver caches unearthed by archaeologists typically contain scrap and cut-up jewelry, both useful for paying by weight. All early denominations of coins were measured in terms of weight, as is the case of the shekel (slightly under half an ounce). This ancient method of denoting value is still with us today in the British pound and the Israeli shekel. Abraham's purchase for 400 shekels of a cave at Machpelah, near Hebron, as burial place for his wife, Sarah, is the earliest biblical mention of a transaction using silver. The final resting place of all the Old Testament patriarchs and their wives, including Isaac and Rebekah, and Jacob and Leah, the cave is a place of pilgrimage for Jews and Muslims alike.

In addition to silver and gold, "utensil money" developed as another medium of exchange. A particular kitchen implement — say a cauldron, spit, or tripod — was assigned a value associated with its use. In the *Odyssey* we learn that Polybus the Theban gives Menelaus "two silver baths, a pair of three-legged cauldrons, and ten talents in gold," testifying to the great value of cauldrons in the first millennium B.C.E. Coinage systems were developed by first setting a standard measure of value by weight or unit, then choosing a valuable metal, such as silver, as that standard.

The world's first coins were struck in the Greek city-state of Lydia during the seventh century B.C.E. Hammering lumps of electrum with engraved dies, minters stamped inscriptions identifying the origin of the coin. In the next century, Croesus, the legendary king of Lydia, substituted pure silver and gold for electrum. Other cities quickly followed suit, including Athens, which minted silver from its mines at Laurion. The owls of Athens, symbols of Athena, the goddess of wisdom and patroness of the city, are perhaps the best known of the deeply imprinted coins of that era. Because of silver's value, coins were not used for daily transactions in the sixth century B.C.E., only for the storage and transport of wealth.

Following the growth of Athens and its democratic institutions, demand for coins as a means of payment for services increased. They paid for everything from jury duty to purchases from local merchants. Consequently, bronze denominations, whose values were a fraction of silver coins, appeared for the first time in the fifth century B.C.E. In the Aegean, the basic unit of weight was the silver drachma (or "handful"), which weighed between three and six grams, probably the weight of a handful of grain. One drachma of silver represented the price of one sheep in the sixth century B.C.E. and the daily salary of an architect a century later. To facilitate trade, the drachma was divided into six silver obols (meaning "skewers"), which trace their origin to earlier utensil money. Obols were still valuable, since a typical worker earned two per day, and so the bronze chalkou was struck. Eight chalkous amounted to one obol.

The gold and silver flowing into Rome from her vast provinces and mines paid for imported luxuries, and was also fashioned into extravagant jewelry and building decorations. But the most significant practical use of gold and silver continued to be as coinage, and Rome's leading exports were the gold aureus and the silver denarius, with twenty-five denarii to an aureus. Toward the end of the first century C.E., the annual salary of a soldier in the Roman Legion was three hundred denarii.

Emperors managed to acquire and squander vast wealth. Tiberius, stepson of Augustus, amassed a fortune estimated at $100 million, which his successor, Caligula, dissipated during a tumultuous four-year reign that ended with his assassination in 41 C.E. Next in line was his uncle, Claudius, who was quite reasonable compared to both his nephew Caligula and his own successor, Nero, who rebuilt Rome following the fire of 64 C.E. Among Nero's projects was the incredible Golden House, decorated with jeweled walls and gold fittings. Nero's extravagance precipitated shortages of precious metals and led in turn to the devaluation of the aureus and denarius, since the shortages were made good by the addition of base metals to gold and silver coins. In fact, during Nero's reign, copper, tin, and lead replaced 20 percent of the silver in the denarius. Once it had started down the path of currency debasement, Rome could not turn back, and by the third century the denarius was 98 percent copper. As the percentage of silver in coins fell, prices in Rome rose to compensate. An inflationary spiral began, with devalued coins being minted in increasing numbers, prices rising, and the value of coins continuing to fall. Roman citizens, desperate for something of real worth in unstable times, quite naturally hoarded silver-rich coins. The same thing happens today. Gold and diamonds are believed to represent safe havens during times of high inflation, while government printing presses churn out increasingly devalued paper currency. Using their debased coins for purchases while hoarding those of pure silver, Romans and their economy provide a textbook example of the so-called Gresham's law, named after and attributed to Henry the Eighth's treasurer and the founder of London's Royal Exchange: "Bad money drives out good."

Toward the end of the third century the weakened Roman Empire split into western and eastern portions, and by 330, Constantine, emperor of the eastern half, moved his capital to Byzantium, renaming it Constantinople. As the western empire crumbled and the ingenuity of former times vanished, the productivity of many Roman mines fell, and they sank into disrepair. Rome doled out its remaining wealth in an increasingly desperate and hopeless attempt to buy protection and peace. Following the sack of Rome in 476 and the collapse of the western empire, coins slowly fell out of use. Europe slipped into the Dark Ages, reduced once again to a barter economy.

The fate of the eastern Roman (or Byzantine) Empire was much different. Gold and silver were still available in abundance there and trade flourished, based on the gold bezant. Byzantium prospered for several hundred years, but finally became enveloped in conflict with her neighboring countries. Nearly a thousand years after the fall of Rome, the fifteenth-century Ottoman Turks swallowed up the remnants of the eastern empire.

Too weak to be fashioned into tools and weapons, gold and silver's value has been wholly determined by the vagaries of human greed. It seems odd indeed, when looked at with a dispassionate eye, that our ancestors

conspired, fought, and died for such fundamentally useless metals. For millennia, the only important role they
played in the development of our world was as currency.
Within the last century, the development of electrical
and photographic applications has given these beautiful
and rare substances new roles.

5

The Age of Iron

And the Lord was with Judah, and he took possession of the hill country, but he could not drive out
the inhabitants of the plain because they had chariots of iron.

— Judges 1:19

IF TURNING TO IRON AFTER GOLD seems like debase-
ment, I should add that iron, while present in high concentrations in the earth's crust, was once so difficult to
find in its pure state it was considered even more precious than gold. However, thanks to the ingenuity of artisans, iron came to play a far more decisive role in
human culture than either gold or silver. Together with
coal, it made possible the industrialization of the modern world.

There are two reasons why the fate of iron turned
out to be so different from that of gold or silver. First, if
you add a small amount of carbon and subject it to a

special heat treatment, iron can be transformed into steel; and second, it is also quite abundant. Once iron and its alloys made the transition from rarities to "working" metals, around the beginning of the first millennium B.C.E., they were the materials of choice when high strength was called for. Without them many of the revolutionary technological innovations of the past five centuries would never have occurred or would have been brought about only with extreme difficulty. The most obvious invention became the impetus behind the Industrial Revolution — the steam engine. Though first fabricated from brass, the steam engine could be mass produced only when it was cast from iron.

Our ancestors' first contact with iron was most likely through meteorites, though their limited number and small size made it difficult to craft iron into anything practical. Occasionally, large meteorites were discovered. For hundreds of years, Eskimos in Greenland fashioned their cutting tools from a 68,000-pound iron-rich meteor. In the 1890s Admiral Robert Peary invested a great deal of time and energy transporting this meteor, as well as others, from Greenland to the American Museum of Natural History in New York City, where they now reside.

In Sumerian, the term for iron was "heaven-metal" and in Egyptian, "black copper from heaven," suggesting that these peoples understood its celestial origin. One reason early inhabitants of the Near East prized iron more highly than gold was because it originated in the heavens, home of the gods. Artisans crafted iron for jewelry and ceremonial weapons. An iron-bladed dagger was a rare and princely gift. Records from Anatolia in the early part of the second millennium B.C.E. tell us that private Assyrian merchants were forbidden to deal in luxury items such as iron, which were reserved for the rulers at Assur. A friendly warning was found on a cuneiform tablet:

> Please do not smuggle anything. If you pass through Timilkia leave your iron which you are bringing through in a friendly house in Timilkia, and leave one of your lads whom you trust, and come through yourself.[9]

Only when metallurgists learned to smelt ores of iron and transform them into steel at the close of the second millennium B.C.E., did iron make the transition from luxury to working metal and did the Iron Age begin.

Most areas bordering the Mediterranean had ample supplies of ores rich in iron, much more so than of copper, which could be found in only a few localities. Iron-rich minerals typically include oxides, such as hematite, which early humans knew as the pigment called red ochre, and magnetite, a magnetic material. Pyrite, more commonly known as fool's gold, is also rich in iron. The Old Testament reminds us how familiar early Near Eastern people were with metal-bearing ores. Moses inspired the long-suffering and down-hearted Hebrews during the Exodus by telling them what a rich land awaited them:

> For the Lord your God is bringing you into a good land, a land of brooks of water, of fountains and springs, flowing forth in valleys and hills, a land of

wheat and barley, of vines and fig trees and pome-
granates, a land of olive trees and honey, a land in
which you will eat bread without scarcity, in which
you will lack nothing, a land whose stones are iron,
and out of whose hills you can dig copper. And you
shall eat and be full, and you shall bless the Lord
your God for the good land he has given to you.
(Deuteronomy 8:7–10)

Moses knew iron and copper would be as enticing to his
followers as milk and honey.

While nature was generous to the smiths of the
Near East, smelting and then heat-treating iron proved
much more difficult than for copper. A simplified description of these processes illustrates the problems.
Mixing chunks of ore, perhaps hematite, together with
charcoal in layers in a furnace (which may have been
only a hole in the ground capped by clay), smelters blew
air in through clay pipes, called tuyeres, to oxidize the
charcoal to carbon monoxide. Rising up through the
layers, the gas reacted with the hematite and reduced it
to iron. Furnaces in antiquity could reach only 1200 degrees, and because pure iron is not molten until 1535 degrees, it was trapped inside the spongy mass called
"bloom iron," consisting of unreacted oxide, unburned
charcoal, and silica impurities. By hammering the bloom
iron at high temperatures, or hot-forging it, blacksmiths
literally squeezed out nearly pure, soft, and ductile
wrought iron.

Iron smelting required care. If the metal was exposed to hot air it would re-oxidize, and if it absorbed

three to four weight-percent carbon, it would form cast iron, which melts at 1130 degrees. Cast iron is brittle and was never used successfully in the ancient Near East. Wrought iron, on the other hand, containing less than 0.1 weight-percent carbon, is ductile, but because it melts at approximately 1500 degrees (a temperature that was not achieved until the late eighteenth century in France) it was impossible to cast. A good question is, therefore, how did early smiths craft soft wrought iron so that it had better properties than bronze, which is stronger than wrought iron? To answer this question we need to look at both the atomic structure and microstructure of iron and its alloys, and how they can be controlled by heat treatment at high temperatures.

Pure solid iron can exist in two different crystal structures: at high temperatures, as a face-centered cubic (FCC) phase called austenite, and at low temperatures, as a body-centered cubic (BCC) phase called ferrite, with atoms at the corners and center of the cube.[10] Both kinds of crystals have large open spaces between the atoms, called interstitial voids — similar in shape to the octahedral configuration of oxygen atoms that build up the structure of clay. Substantial amounts of carbon can dissolve in the austenitic form of iron at temperatures above 910 degrees. Interstitial sites play an important role in this dissolution process. Carbon atoms are relatively small in diameter and can therefore sit in these voids. Actually, the carbon atoms have to squeeze to fit in the BCC unit cell, and the iron atoms surrounding them are slightly displaced from their perfect crystal positions. These local atomic displacements hinder the motion of dislocations and contribute to solid-solution hardening, which increases in magnitude with an increase in the amount of carbon in the iron. An iron alloy containing 0.9 weight-percent carbon is three times stronger than one with 0.1 weight-percent carbon (nine grams instead of one gram of carbon added to nearly 1000 grams of iron) — a remarkable improvement in strength from just a pinch more charcoal.

To substantially strengthen iron, smiths had to form an alloy containing approximately 0.9 weight-percent carbon. This was difficult to do because, as I noted, iron wasn't melted until the end of the eighteenth century; before then it was not possible to form a solid solution by adding carbon directly to molten iron in the way smiths added tin to copper to make bronze. Without understanding what they were doing, blacksmiths solved this problem by heating iron for long periods at high temperatures in a charcoal fire, dissolving carbon in a process called carburization. The carbon atoms move or diffuse into the iron by jumping from one interstitial void to a neighboring one, the way drops of oil or water skip on a hot grill. Ancient smiths never knew that it was the absorption of carbon during prolonged heating that was the key to strengthening iron. They actually thought that they were purifying the iron. Only in the past two hundred years has the role of carbon in strengthening iron been appreciated.

When an iron-carbon alloy with the high-temperature austenite (FCC) structure is slowly cooled to room temperature, it transforms into two coexisting phases: ferrite and a new carbon-rich, iron-carbon compound called cementite, which has a rather compli-cated atomic structure. Ferrite and cementite occur together in a characteristic lamellar-type (platelike) micro-structure called pearlite, because of its pearl-like ap-pearance in the optical microscope. Since cementite is rich in carbon and ferrite is poor in carbon, formation of pearlite from austenite takes time because the carbon atoms must redistribute themselves by a diffusion process.

Pearlite is only slightly stronger than bronze. To produce an extremely hard material from an iron-carbon alloy required rapidly cooling, or quenching, the austenite by plunging it into cold water. Quenching pre-vents austenite from transforming into pearlite; the car-bon atoms have no time to redistribute themselves, and instead a new, highly distorted version of ferrite forms, called martensite, with a structure frozen halfway be-tween FCC and BCC, which is the essence of steel. Dis-locations have such a difficult time moving through martensite that it is five times stronger than wrought iron. I can imagine the joy of the first blacksmith who impatiently plunged hot iron with just the right carbon content into water and was rewarded with steel. Smiths quickly discovered that steel was also brittle and suscep-tible to fracture, and eventually they learned to increase its toughness, or "temper" it, by reheating it to a mod-erate temperature.

My lucky blacksmith likely suffered drastic mood swings: from the euphoria of discovering steel to deep depression when he was unable to reproduce the process. Steel results from smelting ore to iron, hot-forging wrought iron, carburization to just the right composition, quenching, and finally tempering, and any one step can easily go wrong. Today we can only marvel that smiths ever mastered the heat treatments required to form steel, since they could not have had any understanding of why what they were doing worked.

The relatively low temperatures of the furnaces of those days meant that carbon penetrated only very slightly into the iron during carburization. Limited to hardening only thin sections, ingenious blacksmiths in the centuries just before and after the time of Christ fashioned large tools and weapons by stacking individually carburized plates of iron and then hammering them together at high temperatures to form a laminate. Shaping and joining iron must be performed at temperatures above 800 degrees — quite different from copper and bronze, which can be worked at room temperature. To forge red-hot iron, blacksmiths had first to fashion iron tongs, the ancestor of all of our grasping and cutting tools.

All this illustrates why several thousand years passed before iron finally replaced bronze for tools and weapons. Complexities such as having to estimate the temperature from the color of the glowing charcoal made crafting iron implements unpredictable. Where and how did blacksmiths discover these seemingly miraculous processes?

The origins of the revolution in iron metallurgy are still a mystery. Promising locations, based on archaeological evidence, include Cyprus and the Black Sea in the northern part of Anatolia, where there are copper, lead,

and iron ores. Some sand beaches along the Black Sea
are 80 percent magnetite. A famous letter from Hattusilis the Third, king of the Hittites, to Shalmaneser the
First of Assyria, in the mid-thirteenth century B.C.E.,
tells us:

> As for the good iron that you wrote me about, good
> iron in Kizzuwatna in my seal house is not avail-
> able. It is a bad time for producing iron, as I have
> written. They will produce good iron, but so far
> they have not finished. When they have finished, I
> shall send it to you. Today I am having an iron dag-
> ger brought to you.[11]

Archaeologists used to infer from this document that the
Hittites had a monopoly in iron, which is certainly not
true. The letter only discusses a request for iron from
Shalmaneser, one that came at an inconvenient time, and
reveals Hattusilis's attempt to appease him with the gift
of an iron dagger. It also hints at how much time it took
to produce good iron, suggesting that even Hittite
smiths had not mastered the complex processes, and
tells us that in the thirteenth century B.C.E., iron daggers
were still gifts worthy of kings.

Another likely site for the birth of the Iron Age is
Cyprus, where stannite, one of the copper ores containing iron, is found. Knife blades excavated in Cyprus are
compelling evidence that as early as the eleventh or
twelfth century B.C.E., blacksmiths were both carburiz-
ing and hardening steel.

As to how iron smelting was discovered, the most
reasonable answer is that it was a by-product of copper-

ore smelting. A key step in the extraction of iron involves partially oxidizing charcoal to carbon monoxide, a gaseous compound that is critical for the reduction of the ore to metallic iron. All copper ores contain unwanted impurities, concentrated into a residue called "gangue," left over after smelting. Because of its high melting point and high viscosity, gangue held the copper suspended in the form of finely dispersed droplets, essentially useless. Copper smelters overcame this problem by adding a metallic oxide, a "flux," to react with the gangue, lowering its melting point and increasing its fluidity. Denser than gangue, the copper coalesced into large metallic clumps and sank, shielded from oxygen. Serendipitously, the most common flux used for copper ores was hematite, which, we now know, is an iron oxide. The carbon dioxide–carbon monoxide gas mixture from the reaction of charcoal with oxygen during copper smelting can be exactly what is needed to reduce hematite to metallic iron. Indeed, recent excavations have recovered iron particles in the residues from early copper smelting furnaces. The pieces to the puzzle seem to fit together.

Advances in iron metallurgy were accelerated by a number of separate events. Shortages of bronze was one. Massive population movements during the thirteenth and twelfth centuries B.C.E. weakened Egypt and hastened the collapse of the Hittite empire, both major powers in the region. With long-established trade routes disrupted by these migrations, tin supplies were cut off. Archives from the end of the thirteenth century record bronze shortages in Mycenaean Pylos in Greece, where as many as four hundred smiths worked. So desperate was Pylos's plight that, when faced with an invasion, local officials sacrificed religious vessels to the bronzesmith's furnaces to supply arrow and spear heads.[12] Early clay tablets reveal that the supplies of tin were always uncertain, and shortages caused wild fluctuations in price. Assyrian letters, written between 1950 and 1850 B.C.E. in a merchant colony of the city of Assur at Kültepe, located on the Kayseri plain in central Turkey, provide remarkably detailed documentation of the tin and textile trade, including the weight carried by donkeys — 143 pounds — how they were loaded by drovers, and the duties and taxes paid to local princes. We learn that Assyrian merchants marked up imported tin 100 percent and that shipping costs amounted to 10 percent of its value. Tin prices oscillated between six to sixteen shekels of silver for a mina of tin, or one ounce of silver for between four and ten ounces of tin.[13]

The scarcity of bronze may have stimulated the rise of iron, but it was the fall of the Hittite empire at the end of the second millennium B.C.E. that accelerated its spread across the ancient Near East, because blacksmiths from Anatolia fled to the south. Iron metallurgy probably arrived in Palestine with the Philistines who settled along the Canaanite coast (though the quote from the Book of Samuel in my introduction, describing clashes between the Philistines and Hebrews does not mention iron). Other books of the Bible, however, such as Judges, describe how a lack of iron handicapped the Israelites in their battles to occupy Canaan: "Then the people of Israel cried to the Lord for help; for he [the Canaanite king] had nine hundred chariots of iron, and oppressed the people of Israel cruelly for twenty years."

Innovations in the metallurgy of iron had a dramatic effect on its price. In the nineteenth century B.C.E., forty ounces of silver bought one ounce of iron, a ratio of forty to one. By the seventh century B.C.E., technology had advanced so far that one ounce of silver now purchased 2,000 ounces of iron, a ratio of 1 to 2,000. In other words, over a period of one thousand two hundred years the price of iron plummeted by a factor of 80,000 (assuming silver kept a constant value). Today you can buy silver for under five dollars an ounce and steel for 2.5 cents an ounce, so that one ounce of silver now buys slightly less than 200 ounces of iron. Iron would seem to have been cheaper in the seventh century B.C.E. than it is today, but this is because silver is now in much greater supply than in antiquity. This increase in the availability of iron and the consequent dramatic fall in its cost had profound consequences on all aspects of life during the thousand years before the Common Era.

Perhaps the most convincing evidence that smiths in the eighth century B.C.E. strengthened steel by quenching comes to us from Book IX of the *Odyssey,* when Odysseus mentions this process while describing how he and his men blinded Polyphemus the Cyclops with a hot stake.

> Seizing the olive stake, sharp at the tip, they plunged
> it in his eye, and I, perched up above, whirled it

around. As when a man bores shipbeams with a drill, and those below keep it in motion with a strap held by the ends, and steadily it runs; even so we seized the fire-pointed stake, and whirled it in his eye. Blood bubbled round the heated thing. The vapor singed off all his lids on the two sides, and even his brows, as the ball burned and its roots crackled in the flame. As when a smith dips a great axe or adze into cold water, hissing loud to temper it — for that is strength to iron — so hissed his eye about the olive stake.[14]

As Odysseus would have understood, iron's strength was critical to weaponry, and therefore to military campaigns. Assyria entered the Iron Age in the ninth century B.C.E., when Shalmaneser the Third conquered countries that were familiar with ironworking. During his reign, Assyria's treasury overflowed with iron booty and tribute. A storeroom of a later king, Sargon the Second, unearthed at Khorsabad contained 160 tons of ingotlike shapes — for pickaxes, hammers, and plowshares, described by archaeologists as "a wall of iron." Assyria's increased consumption of iron was tied to its military's insatiable appetite for weapons and tools, epitomized by the dictate from Shalmaneser to "sharpen the iron swords that subjugate."[15]

Sargon the Second boasted during his conquest of the city of Ulhu, "The mighty wall, which is made of stone . . . with iron axes and iron daggers I smashed like a pot."[16] With time, the iron sword became a symbol for Assyrian military might. In the eyes of their rulers, Assyrian soldiers were the tool of the god Assur: "With the aid of the iron dagger of Assur, my God, you consumed that whole country with fire." As iron became commonplace it was fashioned into fetters and handcuffs for prisoners. Projecting his immense power, one Assyrian ruler boasted, "I threw him [a conquered king] into bonds and fetters of iron."

The brutality of the Assyrians — they ruled by the proverbial iron fist — made their ultimate destruction a joyous dream for their hard-pressed neighbors. In graphic and grisly detail, the Old Testament Book of Nahum prophesied the fall of Nineveh, their capital:

> The chariots rage in the streets, they rush to and fro through the squares. . . . Woe to the bloody city, all full of lies and booty — no end to the plunder! The crack of whip, and rumble of wheel, galloping horse and bounding chariot! Horsemen charging, flashing sword and glittering spear, hosts of slain, heaps of corpses, dead bodies without end. . . . Wasted is Nineveh; who will bemoan her? (Nahum 2:4, 3:1–3)

And indeed, Nineveh fell to the Medes and Babylonians in 612 B.C.E., its destruction so complete that it was never rebuilt. Within its ruins was buried a wonderful treasure trove for scholars, the library of Assurbanipal, from which nineteenth-century archaeologists recovered the first copy of *The Epic of Gilgamesh.*

The word *steel* is said to come from the Old Teutonic *stah* or *steg,* which means "to be rigid," and even today iron and steel evoke strength. Nazi propaganda regularly invoked the "iron fist" and the Pact of Steel between Mussolini and Hitler to project Fascist military might. In the United States, baseball great Lou Gehrig was called "The Iron Horse," a tribute to his endurance and reliability. And Josef Visarionovich Dzhugashvili would change his name to "Stalin," or "man of steel," and rule the Soviet Union accordingly.

Recognizing the superiority of weapons made of steel, Rome benefited from rich iron deposits on the island of Elba and in Noricum (present-day Austria), and the expertise of sword makers in Toledo. There are stories of how the wrought-iron swords of the Gauls bent during their battles against Roman legions armed with Toledo steel blades, which were "so keen that there is no helmet that cannot be cut by them," as one contemporary account put it. The hapless Gauls had to stop and straighten their blades across their knee after each blow before continuing to fight.

When copper and then iron made the transition from exotic material to working metal, the seeds of a revolution in technology were sown. Artisans could fashion tools that best fit the task at hand, from shaping, sawing, and drilling wood to planting crops and mining ore. Adzes and drills not only had sharper cutting edges than their bronze equivalents, but they retained the sharpness much longer. Iron tools allowed forests to be cleared for farmland in northern Europe, and also transformed the working of the land, leading to large increases in food supply and population. A mere eight pounds of charcoal were required to smelt one pound of iron, as compared to twenty pounds for one pound of copper, a 60-percent drop in charcoal consumption and another advantage of iron over copper. The adverse effects of smelting didn't appear until much later. Although iron took much less wood to smelt than copper, the consumption of iron greatly surpassed that of copper, leading to mining and smelting sites being rapidly stripped of their trees. Gradually, this desolation spread into the surrounding countryside.

Today, the world's production of iron and steel is between one-half to one billion tons annually. A trip to the local hardware store reminds us how we depend upon a wide variety of tools manufactured from steel. Long-nosed pliers, screwdrivers, and saws are still fabricated out of steel — there is simply nothing better and cheaper to use. Ships, bridges, and girders are all made of steel, as are high-precision lathes, milling machines, and drill presses, the tools that make other tools. All manner of vehicles are fabricated out of high-strength steel, as well as the myriad of interchangeable parts for the mass production of appliances. Only within the past hundred years have important new metals emerged — aluminum and nickel, for example. These are starting to play roles approaching those of iron and steel in technology and therefore in our lives.

Iron also played a prominent role in introducing a new material into human history. The metal's high melting point made iron impossible to cast in those early days, but this same characteristic was critical for glassblowing. This revolutionary innovation required hollow tubes resistant to melting that could hold globs of molten glass, a substance with properties remarkably different from any we have so far considered.

6

A Quick History of Glass

> When I was a child, I spake as a child, I understood
> as a child, I thought as a child: but when I became a
> man, I put away childish things.
>
> For now we see through a glass, darkly; but then
> face to face: now I know in part; but then shall I
> know as also I am known.
>
> — 1 Corinthians 13:11–12

STARING AT THE SHATTERED GLASS WINDOW, I fought to control my anger. My oldest son, Adam, had missed his target hitting a tennis ball against our garage door, and shards of glass lay scattered on the floor. Adam reappeared, dustpan and broom in hand, repentant. Taking measurements of the broken windowpane, we went to downtown Ithaca for a replacement. We watched as the owner of the glass shop measured the pane, scribed its surface with a diamond-tipped tool and then, with a quick and deft twist of his wrists, broke it along the line he had scratched. How brittle glass is, that it can break so easily. Glass is never used to carry heavy loads, nor is it expected to survive high-velocity impacts from tennis balls. As a thick plate, glass is weak, but in the form of hair-thin fibers, it is extremely strong, as strong as steel. Such behavior is completely unlike that of metals, whose yield and fracture stresses remain the same whether the component is thick or thin. Pure metals do not shatter. The question is, why this curious dependence on size? And why is glass so brittle?

Glass is actually a supercooled or frozen liquid, similar to water but far more viscous. Indeed, glass's structure is not crystalline but amorphous. It has no particular long-range order, or periodicity, on the atomic scale. A crystal is built up by a unit cell of atoms repeated periodically in three dimensions. Glass, on the other hand, results from atomic-scale building blocks arranged randomly. This difference is well illustrated for silica, SiO_2, a glass-forming substance whose building block is SiO_4, a tetrahedral arrangement of four oxygen atoms with a silicon atom at its center, as we have seen in the structure of clay. A single layer of silica can be crystalline or amorphous. In both forms, neighboring tetrahedra share corners — in the case of glass, in a random network, and in the case of crystals, in a periodic ring pattern. Any amorphous solid, whether silica or metal, is called glass.

Upon heating, glass softens and can then be rolled into sheets, drawn into tubes or rods, cut with a shear, pressed into molds, or blown like a bubble. When it cools it hardens into a crystal-like solid with elastic properties resembling those of ceramics and metals. Depending on how the glass is formulated it can be transparent, translucent, or opaque, colored or colorless. Today glass can also be made so pure that it serves as optical fibers through which information is sent over long distances via pulses of laser light. These fiber-optic cables are quickly replacing copper wires that carry information via a current of electrons. Glass is also lightweight, impermeable to liquids, and easy to clean. Little wonder it has come to play so many roles in our lives.

Any pure crystalline substance, such as copper, salt, or ice, changes from solid to liquid at a specific temperature — known as its melting point — with an accompanying abrupt change in volume. Water's transition from solid to liquid illustrates this volume change. Ice floats because the water molecule occupies a larger volume in the solid than it does in the liquid state. Since it goes from a supercooled liquid to a so-called normal liquid, glass goes through a very different kind of transition from solid to liquid: its volume changes continuously with temperature, as does its stiffness and viscosity. This behavior was crucial for early artisans, who sought new ways to shape glass rapidly without it cracking.

Whether or not silica glass forms on cooling depends on the ability of the silica tetrahedra to order themselves, and that in turn depends on how quickly the liquid is cooled. For example, when cooled rapidly, molten silica forms a glass called "fused silica." When cooled extremely slowly under the right conditions, it can solidify into one of several crystal structures, including quartz, which is stable at low temperatures, or cristobalite, with a cubic unit cell similar to that of diamond and stable at high temperatures. Quartz and cristobalite crystals are both built up by the periodic stacking of SiO_4 tetrahedra to form three-dimensional network structures. (For the reader puzzling over how the stacking of SiO_4 tetrahedra can give a structure with composition SiO_2, in three-dimensional structures the tetrahedra are arrayed tip-to-tip, with the oxygen atoms at each corner shared by two tetrahedra. One-half of the oxygen atom at each corner is counted as belonging to any single tetrahedron, so that the number of oxygen atoms associated with one tetrahedron is computed as 4 corners times ½ oxygen atom per corner equals 2. Two oxygen atoms at corners plus one silicon atom at the center of the tetrahedron give the composition SiO_2.)

Because silica structures contain large amounts of open space, they also have low density, or weight per volume. An ordered structure is always more stable than a disordered structure of the same composition, and after a long period of time at low temperatures, glass eventually crystallizes in a process called devitrification. Glass is one of many technologically important materials that are metastable; that is, not in their most stable form but still quite useful. Incredibly, diamond and martensite in steel are other examples of metastable substances. Were there a heavenly switch that could suddenly transform metastable structures into stable structures, and some divine hand threw it, that beautiful blue-white diamond in the showcase at Tiffany's would turn into a chunk of graphite, otherwise known as pencil lead. All our glass windows would become opaque and shatter. Whole cities would collapse, because steel girders would buckle under the weight of skyscrapers. What would be left is an incredible mess, so may we stay ever metastable. We are indeed fortunate that glass needs thousands of years to crystallize at room temperature, remaining glassy over many human lifetimes.

How and when did glass enter into human history? Obsidian, as we've seen, is a naturally occurring black translucent glass that forms when silica melts deep within the earth, then cools and solidifies after being forced to the surface by volcanic eruption. Early humans cleaved obsidian for tools and weapons, and it was crucial for Çatal Hüyük's trade in the seventh millennium B.C.E. Because pure silica melts at 1713 degrees, a temperature not attained under controlled conditions until the late eighteenth century, sodium carbonate, known to Egyptians as natron, was added to lower its melting point. Egyptian worshipers purified themselves by chewing natron (the first toothpaste) or rinsing in a solution containing it. Egyptian priests embalmed mummies with natron. While silica was widely available in the form of sand, the largest source of the natron needed for glassmaking was an oasis in the Western Desert adjacent to the Nile delta, known even today as Wadi Natrun. During the Nile's flood season, water seeps into this oasis and forms small pools, which dry up during the hot summer, leaving behind deposits of sodium carbonate and sodium bicarbonate (baking soda), both critical to glassmaking. Along the Syria-Palestine coast known as the Levant, ashes of plants and trees provided similar compounds, rich in potassium instead of sodium.

Stephen L. Sass

Sodium silicate glass results when silica reacts with natron. While pure silica is solid until 1713 degrees, natron melts at 850 degrees and then reacts violently with the silica to produce carbon dioxide. The sodium oxide contributed by the natron is a "network modifier," breaking up the continuous network of silica tetrahedra and lowering its viscosity and melting point from that of pure silica, allowing a more disordered structure to form. The problem was that sodium silicate glass deteriorated so badly when exposed to water that it was essentially useless. To improve its resistance to water, early glassworkers learned to add lime, calcium oxide. Both calcium and sodium atoms occupy random positions throughout glass's network structure. Calcium strengthens the network against weathering because the chemical bond between calcium and oxygen atoms is 50 percent stronger than that between sodium and oxygen atoms.

As to who first made glass, all the ancient evidence points to one locale: the Mediterranean littoral stretching from Mount Carmel in the south to Tarsus in the north, known as the Canaanite-Phoenician coast. Phoenicians were Canaanites who settled along the seacoast, built boats, and bartered locally before becoming long-distance merchant traders. As early as 3000 B.C.E., the coast served as the land bridge between Mesopotamia, a hotbed of innovation, and Egypt, a large market for goods, ideas, and artisans. While it is impossible to be certain, the oldest glassmaking locales seem to have been located at the mouth of the Belus, a small river just north of Mount Carmel and present-day Haifa. Pliny's *Natural*

103

History, written in the first century C.E., relates what was common knowledge to Greek merchants frequenting that area for many centuries before his time, and contains the best evidence linking glassmaking to this region of the Mediterranean:

> That part of Syria which is known as Phoenicia and borders on Judaea contains a swamp called Candelia on the lower slopes of Mount Carmel. This is believed to be the source of the River Belus, which, after traversing a distance of five miles, flows into the sea near the colony of Ptolemais [Akko, present day Acre]. Its current is sluggish and its waters are unwholesome to drink, although they are regarded as holy for ritual purposes.
>
> The river is muddy and flows in a deep channel, revealing its sands only when the tide ebbs. For it is not until they have been tossed by the waves and cleansed of impurities that they glisten. Moreover, it is only at that moment, when they are thought to be affected by the sharp, astringent properties of brine, that they become fit for use. The beach stretches for not more than half a mile, and yet for many centuries the production of glass depended on this area alone.
>
> There is a story that once a ship belonging to some traders in natural soda [natron] put in there and that they scattered along the shore to prepare a meal. Since, however, no stones suitable for supporting their cauldrons were forthcoming, they rested them on lumps of soda from their cargo. When these became heated and were completely mingled with the sand on the beach, a

Stephen L. Sass

strange translucent liquid [*novi liquoris*] flowed forth in streams; and this, it is said, was the origin of glass.[17]

Sand from this area is approximately 80 percent silica and 9 percent lime, so all that was needed for good-quality glass was natron and a hot fire, which the traders apparently provided.

Pliny's tale, while appealing, is, alas, not credible. Campfires cannot reach the temperatures in excess of 850 degrees required for the glass-forming reaction. Early glasses likely had a very different origin. In the fourth millennium B.C.E. and possibly even earlier, faience, another glasslike substance, used for coating clay vessels, was formulated by heating a mixture of silica, natron, and lime until the natron melted and coated the silica grains. On cooling, the liquid natron, having reacted slightly with the silica and lime, solidified into a thin layer of glass, bonding the mixture of silica grains together. Glass was probably first formed after faience workers raised their furnace temperatures, allowing the natron to strongly react with the silica and completely dissolve it.

Although the glass industry is believed to have its origin in the Near East during the sixteenth century B.C.E., it is more likely that this was when glassworkers made a crucial improvement in their technology and crafted something really useful out of the material — containers. Before then, glass was used mainly in glazes on clay pots, dating from the fourth millennium B.C.E., and on small trinkets such as beads, from the beginning of the third millennium B.C.E. As with metals, the earliest uses of glass were decorative. Glass came to play an important role in the lives of the common people only when relatively large quantities of vessels could be fashioned economically.

By the sixteenth century B.C.E., Mesopotamian and Levantine glassworkers had refined the techniques and tools needed to handle softened glass and could now coil it around a core to make containers. Artisans first built up the shape of the interior of a vessel by packing a thick layer of clay and dung around a metal rod. Thin strands of glass drawn from a molten chunk of glass were then wound around this core. Following heating, workers smoothed its exterior surface by rolling it on a stone slab, a process known as "marvering." They decorated the vessel by combing colored threads of glass in a variety of patterns on its surface and flattening them, again by marvering. Finally, after adding handles and a base and cooling the container, the glassworker broke and removed the core. This "core-forming" technique was time-consuming and only capable of producing small containers three to four inches high, and therefore not suitable for creating storage and drinking vessels. From the standpoint of efficiency, shaping glass over a mold was a much more important process for plaques, pendants, cups, and bowls. More than a thousand years were to pass before artisans finally arrived at a simple way to fabricate glass bottles rapidly.

Bowls were also made by stacking glass disks. Artisans fused together groups of thin, individually colored sticks of glass arranged in a pattern to form rods, which they then stretched to reduce their diameter. Cutting the rods into thin disks, glassworkers stacked them on a core in the shape of the bowl, with an outer mold added to hold the disks in place as they were fused in a kiln. Polishing the inner and outer surfaces produced particularly beautiful mosaic or milleflori bowls.

Following Pharaoh Tuthmosis the Third's conquest of Phoenicia and Syria in the early fifteenth century B.C.E., glass workshops that utilized the core-forming method appeared in Egypt. An inscription in the temple of Amon-Re at Karnak, recording Tuthmosis's great victory at Megiddo in 1478 B.C.E., includes a molten stone on his list of booty. Most likely it was a piece of translucent glass. The inscription is the first written evidence of glass along the Levantine coast. Beautifully colored vases and cups have been recovered from Egyptian graves of that era, including the Tuthmosis tomb.

Large Canaanite-Phoenician fleets operating out of Sidon, Tyre, and Byblos, north of Akko, made the area around the River Belus the main focus of Mediterranean trade routes in the latter part of the second millennium B.C.E. Traveling merchants carried tales of glassmaking and glassworkers to Cyprus and the Aegean in the fourteenth and thirteenth centuries B.C.E. These Phoenicians and their neighbors, the Aramaeans, were also responsible for exporting a discovery that proved even more vital to human culture than glassmaking. It is worth a quick aside.

Responding to the demand for a convenient means of recording commercial transactions during the second millennium B.C.E., Canaanite scribes invented the alphabetic approach to language. There are three fundamentally different ways that languages can be written. The first, the ideographic, uses a separate sign, an ideogram or pictogram, for each word or idea, and requires hundreds if not thousands of different signs for the expression of complex thoughts. The second, the syllabic, has a separate sign for each syllable, a group of letters including a vowel that makes a specific sound: it typically needs up to one hundred different signs for the articulation of involved ideas. The third system of writing, the alphabetic, uses specific signs for the main sounds in a language and typically requires less than fifty signs, which together comprise an alphabet. At the end of the second millennium B.C.E., the Phoenicians devised the most successful alphabet, which was adopted by the Aramaeans and used as the standard form of writing by the Assyrians and Persians. The Phoenician alphabet made its way westward and was favored by the Greeks and Romans. All Western alphabets can be traced either to Phoenician or Aramaean scripts.

Glassmaking along the Canaanite-Phoenician coast and in Mesopotamia flourished until about 1200 B.C.E., then went into decline for three centuries, and even longer in Egypt. Archaeologists have recovered few glass artifacts from this period. This hiatus was linked to the turmoil enveloping the entire area, including the collapse of the Hittite empire and the eclipse of Egypt following a series of debilitating attacks from the Sea Peoples, who were thought to have migrated from farther west in the Mediterranean. A dark age descended on the eastern Mediterranean, disrupting trade and craftwork.

Important aspects of our spiritual and technological world today owe much to these upheavals in the eastern Mediterranean. Shortages of tin from disruptions in trade inflated the price of bronze, stimulating innovative smiths to transform iron from a rarity to a working metal. Another consequence of the turmoil was a political vacuum that allowed small states, which under normal circumstances were vassals of major powers, to find their own way undisturbed, at least for a time. By far the best known of the nations emerging was Israel. Fortuitously, or perhaps by divine intervention, it was then that Moses was leading the Hebrews across the Sinai Desert to the land of Canaan. Unhindered by their once-powerful neighbors, the Israelites were free to pursue their destiny for several hundred years. Indeed, events in the Near East at the transition between the Bronze and Iron Ages have played a crucial role in shaping Western civilization.

Also during this period, one group of the Sea Peoples, the Philistines, invaded the Canaanite coastal plain south of Mount Carmel, established the Pentapolis, the five city-states of Ashdod, Ashkelon, Gaza, Gath, and Ekron, and gave a name — Palestine — to the land. A short time later, the twelve tribes of Israel, crossing the Jordan River from the east, established themselves in the hill country inland from the coast. Late in the eleventh century B.C.E., the Israelites came under pressure from their neighbors, particularly the Philistines, who overran Shiloh, the religious center of the Israelite faith, and carried away the Ark of the Covenant. Uniting under their first king, Saul, the Israelites expelled the Philistines from the hill country. Continuing their campaign, Saul and his son Jonathan attacked the Philistines in the Plains of Esdraelon, where they were defeated and killed, inspiring David to compose his eloquent elegy over their bodies, described at the beginning of the second Book of Samuel.

With David as their king, the Israelites subdued the Pentapolis and formed a United Kingdom. For a few decades under David and later his son, Solomon, Israel was the most powerful and prosperous state in the Levant. Unfortunately for the Israelites, during this period and particularly in this locale, the crossroads of the Near East, long-term tranquillity was elusive. The unity of a state was tied to the strength of its leader and, following the death of Solomon, the Israelite nation split in two — Israel in the north, and Judah in the south. In addition to arguing between themselves, both kingdoms were soon exposed to the wrath of their neighbors, the resurgent Egyptians to the south, the Aramaeans to the north, and worst of all, the Assyrians to the northeast. Following seemingly endless conflicts during the ninth and eighth centuries B.C.E., the Northern Kingdom of Israel, its capital Samaria besieged by the Assyrian king Shalmaneser V, finally fell after being sacked in 722 B.C.E. by his successor, Sargon the Second.

Despite, or perhaps because of, these tumultuous events, glassmaking revived along the Phoenician coast and in Mesopotamia during the early years of the first millennium B.C.E. Evidence of this renaissance includes the famous cup with the inscription of Sargon the Second cut from a block of glass, now in the British Museum. According to an apocryphal tale, during her historic visit to Solomon, the Queen of Sheba saw a pool of water in his palace and raised her skirts to cross it, revealing her (shapely) legs to him. Smiling, Solomon told her that the palace was tiled with glass.[18]

The rebirth of the glass industry in Mesopotamia was stimulated by the military campaigns of the Assyrians, who forcibly imported captive artisans to help decorate their capital. Plagued by the perennial problem of fielding a large army whose soldiers fought instead of tending and harvesting crops, Assyrian and Babylonian kings satisfied their agricultural needs by deporting entire populations of vanquished countries. Following his brutal capture of Jerusalem and destruction of Solomon's temple in 587 B.C.E., Nebuchadnezzar, the Babylonian successor to the Assyrians, sent the Israelites into their Babylonian exile. Within fifty years, Babylon in turn capitulated to the Persians under Cyrus the Great. Establishing a commonwealth of provinces, enlightened Persian rulers such as Cyrus and then Darius encouraged the Israelites to return to Jerusalem and rebuild their temple, as we hear in Ezra:

> Thus says Cyrus king of Persia: "The Lord, the God of heaven, has given me all the kingdoms of the earth, and he has charged me to build him a house at Jerusalem, which is in Judah. Whoever is among you of all his people, may his God be with him, and let him go up to Jerusalem, which is in Judah, and

rebuild the house of the Lord, the God of Israel —
he is the God who is in Jerusalem."

Returning home from Babylon, the Hebrews put
the expertise they acquired during their exile in the cul-
tural and technological center of the world to good use.
And so, ample evidence of glassmaking was to be found
in Galilee from just before the start of the Common Era.
Perhaps most extraordinary was the discovery of an
eight-ton slab of raw glass, measuring 11 feet by 6¼ feet
by 1½ feet thick, at the bottom of a cistern at Beit
Shearim.[19] It is still not clear for what this enormous
chunk of raspberry-colored glass with greenish streaks
was going to be used. It had not been moved from where
it was cast more than two thousand years ago, and likely
represents the failed attempt of some early entrepreneur
to provide chunks of raw glass to artisans, who would
then remelt and work them. Very few pieces of glass so
large exist even today, aside from the two optical mirrors cast by the Corning Glassworks for the 200-inch
telescope at Mount Palomar.

The history just traced emphasizes events along the
Levantine coast and in Mesopotamia, where innovations in glass forming are believed to have originated.
But Egypt also had an advanced glass industry, as testi-
fied by her beautiful fifteenth- and fourteenth-century
B.C.E. core-formed vessels. In fact, the most detailed
information on glassmaking in the mid-second millennium B.C.E. comes from workshops excavated at El-
Amarna and Malkate in Egypt, where quartz pebbles,
pigments, and glass rods have been recovered.

Exchanges between different parts of the Near East fostered innovations in glassworking, just as earlier interchanges had stimulated the development of domesticated wheat, animal husbandry, and the tools and techniques needed for large-scale farming. During the sixth or seventh century B.C.E., high-temperature furnaces were developed in Mesopotamia, making the molten glass (the melt) more liquid. Cuneiform tablets in Assurbanipal's library at Nineveh, dating from the seventh century B.C.E., that contain information about medicine, chemistry, geology, and alchemy also describe a sophisticated glassmaking industry, including high-temperature kilns. Artisans were learning to estimate temperatures by the color of the molten glass. According to the tablets the glass would "glow red"; "glow green/yellow"; and "glow golden yellow," with the latter indicating the highest temperature.[20]

These improved furnaces led to glassblowing, a discovery that rapidly transformed glass from a luxury into an everyday item. Returning to Jerusalem from their exile in Babylon, the Jews may have been instrumental in stimulating this innovation, which was first practiced at Sidon in the first century B.C.E. That would mean five hundred years passed from the development of hotter kilns to when ingenious artisans first began to blow glass. This interval is a good estimate of the incubation period for technological advances two thousand years ago. In our world, the time between discovery and large-scale industrial production has shortened dramatically, and frequently is measured in only a very few years.

With hotter furnaces, an ancient glassworker could

now pick up a molten glob, or "gather," of glass at the end of a three- to five-foot iron blowpipe, and inflate it to form a bubble, which he then rapidly shaped into a bottle. If the object was simple, the whole process might take only a few minutes. This technique was not practicable until glass could be made hot enough to flow easily under the pressure of air from the artisan's lungs. Constantly rotating the blowpipe to maintain the symmetry of the molten bubble and to prevent it from sagging, he rolled the glass on stone, shaped it with wooden paddles and pincers, and finally severed it with shears. Because glass undergoes a continuous change of structure on solidifying, the bubble could be cooled and reheated many times without cracking. Glass remains amorphous and can survive such thermal cycles without shattering, because the volume change between the solid and the viscous liquid occurs gradually with temperature. Of course, if a glass object were heated or cooled *rapidly* — for example, by quenching, as with steel — it would shatter, because the large difference in temperature (and interatomic spacing) across the thickness of the glass would generate high internal stresses.

Glassblowers transformed the hot sphere into a teardrop shape either by a swinging motion or by allowing the bubble to sag under its own weight. Once the piece was shaped, the glassworker attached its base to a metal rod called a pontil, knocked it off the blowpipe, and then added a rim and handles. Stresses inevitably built up in the glass from uneven temperature distributions during blowing, and to relieve these the artisan reheated and then slowly cooled the vessel to room temperature over a period of one day. Glassblowing techniques developed at the time of Christ remain fundamentally unchanged today. The high melting point of iron, which had been a problem for smelters and blacksmiths, was now an advantage, for it meant blowpipes made from it could safely be dipped into molten glass to pick up a gather.

Artisans also blew glass into molds, allowing them to rapidly form containers whose shape, size, and decoration were reproducible. All they needed was a two-piece mold of the bottle or vase, hinged at one end. Air from the glassblower's lungs forced the molten bubble to conform to the interior of the mold. His breath, in a sense, replaced the core in the core-forming process. Cheap glass tableware, storage bottles, and vases were now widely available, and before long displaced their pottery counterparts. The Roman square bottle that appeared about 40 C.E. became the prototype of all future glass bottles and a mainstay of the ancient world. Prior to the first century B.C.E., artisans crafted containers for liquids by grinding stone or glass, or by first shaping clay on a wheel, then carefully drying and firing, as well as by the core- and mold-forming processes discussed earlier. Glassblowing replaced all these laborious processes with a procedure that was simple and elegant. In addition to being easier to form, containers of glass were also more sanitary. Unglazed clay tended to pick up odors and small particles whereas glass did not.

By the start of the Common Era, glass, once so precious that artisans used it to decorate the ceremonial shields of Macedonian kings and the golden mask of

Tutankhamen, could now be produced with ease. Vast numbers of vessels were available to broad levels of society during Roman times. Besides bottles, popular favorites included cups shaped like human heads, or decorated with scenes from the Roman circus, showing favorite chariot teams and gladiators — souvenirs of a day at the Coliseum. So proud of their handiwork were glassworkers such as Artas, Philippos, and Ennion that they frequently engraved their names on the clay molds used for blowing. They were the Tiffany, Lalique, and Steuben of their day. Ennion seems to have been particularly well known in the first half of the first century C.E. He traded his wares across the Roman world, shipping mold-blown cups, beakers, and two-handled bottles from his factories in Syria and Italy.

The first century of the Common Era saw rapid advances in technical know-how, with ingenious artisans blowing panes of glass by piercing the bubble and then spinning it to flatten the molten liquid. A telltale knob remains where the plate was joined to the blowpipe. The first glass windows appeared at the time of Christ. Glassblowers who followed the Roman legions pushing to the north and the west, up the Saône, Rhone, and Rhine Rivers, deep into the heart of Europe, spread these innovations. As a consequence, during the second century, glass factories sprang up in new towns in Gaul, Britannia, and, with Colonia Agrippinensis (Cologne) as its center, the Rhineland. A thousand years later, the expertise found in these areas would craft the glorious stained glass adorning so many medieval cathedrals.

Though the great quality of glass is its transparency, early glass was opaque. Bubbles or small particles trapped in the viscous melt scattered light strongly. It was also colored from impurities, such as iron. During the first millennium before the Common Era, glassworkers had discovered the art of controlling color by adding metallic oxides to the melt. With this, glass chemistry became more complex. Additions of copper or iron to the molten liquid produced different effects, depending upon the atmosphere in which the glass was heated. When heated under the neutral or oxidizing conditions typical of air, glass containing five to twenty weight-percent copper had a transparent blue color. When held at high temperatures in a reducing atmosphere — that is, with little oxygen present — the same glass became an opaque bright red, since under these conditions the copper exists in the form of small particles of red cuprous oxide. Thus, blue glass containing copper becomes red on heating either in a closed container or in the presence of charcoal, which limits the amount of oxygen in the environment. The magnificent stained-glass windows in Chartres, Sainte-Chapelle, Cologne, and many other medieval cathedrals of Europe owe their striking colors to knowledge of such chemistry.

Seeking to satisfy connoisseurs who avidly sought colorless rock crystal (quartz) plates and vases, artisans learned to bleach out impurities such as iron with antimony oxide. Colorless glass used to command premium prices in the marketplace. Today, it is commonplace.

Why does adding antimony oxide form colorless glass? Atoms in metals are neutral (uncharged) because they have the same number of electrons (negatively charged) as protons (positively charged), which cancel each other out. Certain metal atoms can be present in chemical compounds with different numbers of electrons in their vicinity. Iron is a good example, as illustrated by its ability to form two different compounds with oxygen, FeO and Fe_2O_3. In FeO, iron exists as Fe^{2+}, which is doubly positively charged because it is missing two electrons. An atom with an electric charge is called an ion. Positively charged iron ions are attracted to negatively charged oxygen ions and react with them to form a chemical compound. Since the double positive charge of one. Fe^{2+} ion is balanced by the double negative charge of one O^{2-} ion, the neutral compound that forms is written FeO. In Fe_2O_3, the iron exists as Fe^{3+} ions. The Fe^{2+} ions can be transformed into Fe^{3+} ions by changing the amount of oxygen present in the atmosphere.

Adding antimony oxide to glass is the equivalent to adding a high concentration of oxygen. The antimony makes it colorless, because how an atom absorbs light depends upon the number of electrons in its vicinity. The electrons absorb the light and then reemit it by oscillating rapidly. Blue light has a wavelength of 4,500 angstroms (remembering that one angstrom is one ten-billionth of a meter), while red light has a wavelength of 6,500 Å, with the visible spectrum ranging from 4,000 to 7,000 Å. White light contains all the visible wavelengths and their corresponding colors are present equally. If a metallic impurity in glass absorbs one particular wavelength in the impinging white light more than others, it subtracts that wavelength. The result is that the color of the light either passing through or being reflected by glass now consists of only the remaining wavelengths. Adding antimony oxide transforms ubiquitous iron impurities from the ferrous (Fe^{2+}) to the ferric (Fe^{3+}) state. Since Fe^{2+} ions have one more electron than Fe^{3+} ions, they also absorb light differently. The color of the glass changes from pale green to pale yellow, which we see as colorless.

As I've said, opacity in glass occurs when bubbles and small particles scatter light; their ability to do this depends on the incident wavelength and the spacing and size of the particles. When there are large numbers of small, closely spaced particles present, none of the impinging light is transmitted, and the glass is opaque. Thus glass containing fine particles of cuprous oxide can be both red and opaque.

Once they discovered how to control color, imaginative glassworkers crafted extraordinary works of art, epitomized by the Rothschild Lycurgus Cup from the fourth century C.E., now displayed in the British Museum. The cup is named for Lycurgus, the king of Thrace, who, legend has it, so enraged Dionysius that the god placed a curse on Thrace. The cup depicts Lycurgus being strangled by grapevines. Small amounts of gold and silver dissolved in the glass make the cup jade green in reflected light and magenta in transmitted light. Fashioning such a magnificent cut-glass cup would be considered an extraordinary achievement today; given the crude tools available to glassworkers just after the time of Christ it is even more remarkable. Techniques for cutting, grinding, and polishing stone were well suited for glass. The Lycurgus Cup, an example of a "cage cup," or *diatreta,* testifies to the extraordinary skill possessed by glassworkers of the times. Casting or blowing a thick-walled cup, the artisan would then grind away the material between a thin outer layer of glass and the bulk of the cup, leaving a delicate network of glass circles.

From the time of the Roman Empire to the end of the thirteenth century, approximately a thousand years, glass was used only for containers, tableware, and windowpanes. The earliest scientific application of glass came when lenses were first ground for spectacles in the late 1200s, and then for optical telescopes and microscopes in the late 1500s and early 1600s, which opened up new worlds to the observant eye. Telescopes permitted Galileo to follow the movement of the planets around the sun. Microscopes enabled van Leeuwenhoek to discover bacteria. Today, glass appears as fibers for communication using laser light (the conduit of the "electronic superhighway"), as a strengthener for fiberglass composites, and as the transparent envelopes of television tubes.

Thus far I have dwelt on glass's desirable properties, but obviously it also breaks easily. In this regard components fabricated from metal have an enormous advantage over glass. Their failure is often (though as we saw with metal fatigue, not always) preceded by extensive plastic deformation, giving ample warning before fracture. Glass fails catastrophically and without warning because of the presence of surface flaws —

typically, cracks with very sharp tips — where the local stress is multiplied many times relative to the applied stress. The longer the crack, the smaller the load needed to increase its length. Introduced either during the glass-forming process or by subsequent handling, cracks dictate the fracture strength of glass.

The presence of cracks leads to an explanation for the puzzling and unique behavior described at the beginning of this chapter. Hair-thin glass fibers are much stronger than thick ones because as the dimension of the glass increases, the likelihood of having a long (and more potent) crack also increases. Since thin fibers can by their nature contain only relatively short cracks, which require large stresses to propagate, they are stronger than thick fibers.

To cleave glass plates to size, glaziers score them with a diamond-tipped tool, like the one Adam and I saw the glazier use at the store. This extreme sensitivity to flaws is the source of considerable worry to aircraft designers, which is why they place transparent plastic shields between the windows and the cabin of a jet airliner. This is not done to protect passengers, as you might have thought, but rather to keep them and their diamond rings away. The plastic also keeps body oils off the glass, since organic greases frequently accelerate crack growth. Metals also contain cracks, but local plastic deformation involving dislocation motion relieves the high stresses at their tips before they can grow any longer. Since dislocations do not exist in glass, they cannot relax the high stresses at the tips of cracks, which are free to propagate catastrophically: glass shatters.

Understanding the relationship between cracks and brittleness can be used to improve the resistance of glass to fracture or, in the jargon of materials scientists, improve its fracture toughness. A stress that pulls the two sides of a crack apart, a *tensile* stress, is the most destructive to glass. Conversely, the introduction of *compressive* stresses, which push the sides of the cracks back together and cause them to close up, improves glass's resistance to fracture. Compressive stresses can be introduced into the surface of glass by ensuring during cooling that the interior of the glass contracts more than the flaw-ridden surface. One way of doing this is by a judicious choice of chemical components, because the amount glass contracts on cooling depends on its composition. If you sandwich a layer of glass between two outer layers that undergo a smaller contraction upon cooling, you will find the interior of the glass shrinks more than the outer layer, putting the exterior surface into compression. This is one of several ways that tempered (or bulletproof) glass windows to protect presidents and popes are made.

In an apocryphal tale, a glassworker invented shatterproof glass at the start of the Common Era. The intrepid inventor then demonstrated his discovery to Emperor Tiberius by dropping a glass bowl, which only dented. He pounded out the dent with his hammer. (Glass like this does not exist today, and probably never did.) A stunned Tiberius asked the ingenious artisan whether he had told anyone else about his discovery. Hearing that he had not, Tiberius quickly had the hapless inventor's head chopped off, supposedly to protect the value of his own glass collection. Alas, a remarkable invention was lost.[21]

First iron and then glass made the transition from precious to commonplace during the first millennium B.C.E. A clear lesson can be drawn from this. Materials make a significant impact on human society only in proportion to how economical they are to obtain and then fashion into something useful.

During our century, glass began to be used in new ways, in particular for glass fibers for communications and in composites. Returning to the time of Christ, we should note that many of glass's constituents, such as sand and lime, were also crucial for the Romans' monumental building projects. Silica, for example, was a component of that landmark Roman invention, concrete, a construction material as important today as it was two millennia ago. A reason to turn now from small-scale materials applications to those with grander dimensions.

7

Building for the Ages

> In Uruk [Gilgamesh] built walls, a great rampart, and
> the temple of blessed Eanna for the god of the firma-
> ment Anu, and for Ishtar the goddess of love. Look at
> it still today: the outer wall where the cornice runs, it
> shines with the brilliance of copper; and the inner
> wall, it has no equal. Touch the threshold, it is an-
> cient. Approach Eanna the dwelling of Ishtar, our
> lady of love and war, the like of which no latter-day
> king, no man alive can equal. Climb upon the wall of
> Uruk; walk along it, I say; regard the foundation ter-
> race and examine the masonry: is it not burnt brick
> and good. The seven sages laid the foundations.[22]
>
> — *The Epic of Gilgamesh*

MY SON ERIK AND I were sprawled out on the side-
walk of the Commons, a pedestrian mall that runs
through the center of downtown Ithaca, listening to a
summer jazz concert. I spent the time casually surveying
the buildings around me. The vista of Main Street in

small-town America: rows of tan, red, and white brick-framed three-, four-, and five-story buildings ran down one side of the mall, facing a redbrick-and-concrete-fronted department store. Many of the materials used for their construction have been unchanged for at least two thousand years. The bricks are essentially the same the Sumerians employed for their ziggurats. And Roman engineers invented concrete for their grandiose constructions, shortly before the start of the Common Era.

What if concrete and steel did not exist? A narrow eleven-story building rises over the downtown of Ithaca. It certainly could not have been built from stone and wood alone. Without the steel used in the skeletons of its skyscrapers and spans of its bridges, a city like New York would have developed on a much smaller scale. Materials scientists and our predecessors helped lay the foundations of modern cities.

Human beings build for protection, comfort, storage, to honor our divinities, or to glorify ourselves. We erected our earliest substantial structures by tamping walls of earth into place. Stacking sun-dried mud bricks was another early approach; each level was covered with mud, which produced a solid clay wall upon drying. The frequent occurrence of mounds (or "tells") throughout the Near East spoke both to the prevalence and short life of mud brick construction. Even in such an arid climate, city-dwellers were continually forced to rebuild their constantly eroding homes. Egyptians erected pyramids by stacking cut stones, while Mycenaeans fashioned domes by forcing small wedges between the stones to lock them into place.

Sumerians used mud bricks to wonderful effect for their monumental ziggurats. Stacking sunbaked bricks for a core, builders then covered them with several layers of "burnt bricks," fired in a kiln as a shield from the infrequent but destructive torrential rains of Mesopotamia. Cementing the outer layers together, bitumen mortar protected the internal mud bricks with a waterproof coating. Drains in the massive walls channeled away seepage from the terraced gardens above. High over the hot dry Mesopotamian plain the ziggurats, with their lush greenery, were dazzling vistas for the people of Sumer going about their daily chores. More than a thousand years later, the Hanging Gardens of Babylon, descendants of these early ziggurats, were one of the Seven Wonders of the Ancient World.

Using bitumen was an extraordinary innovation. Today a petroleum product that is the viscous residue of oil refineries, and typically used in the form of asphalt for highways and roof coatings, bitumen was the superglue of the third millennium B.C.E. It could be found in seepages near the town of Hit ("bitumen" in Akkadian) on the Euphrates River north of Babylon, ninety miles west of present-day Baghdad. Bitumen bonded well to the porous Sumerian bricks fired at the relatively low temperatures of 550 to 600 degrees. Sumerians thought of bitumen as excrement from the seething underworld, since it was typically found together with vile-smelling natural gas — their cuneiform symbol for bitumen consists of two signs meaning "well" and "abyss." Bitumen came to represent evil spirits and is likely the inspiration for the biblical description, "lake of pitch which is hell." In antiquity, natural gas was thought to be the origin of omens from the gods. The Assyrian king Tukulti Ninurta wrote at the beginning of the ninth century B.C.E.: "Opposite Hit, close to the sources of bitumen, I camped at the place where the voice of the gods issueth from the Usmeta rocks."

Lighter, less viscous components of the seepage, called naphtha then and gasoline today, were so flammable that to the ancient world they were nuisances. Priests frequently based their prophecies on patterns they read in oil poured in water. Cunning oracles, in fact, controlled these figures (and their auguries) by judicious choices of oil. The Greek *naphtha* can be traced back to the Babylonian *naptu* used to describe means of prophesying in 2000 B.C.E. Seepages of bitumen and naphtha at Hit were early hints of the vast reservoirs of petroleum underlying much of Mesopotamia, today the primary source of wealth for the countries bordering the Persian Gulf.

Gathering bitumen was a relatively easy task. It floated on the waters of the Dead Sea, where traders from Jericho in the eighth millennium B.C.E. could pull it into their boats because of its molasseslike consistency. After drying the bitumen, they chopped it into smaller chunks for trade. In addition to waterproofing and joining bricks together, bitumen was used to cement ax heads to wooden handles and to set colored stones in jewelry.

Clay, lime, and gypsum (calcium sulfate) were

among the earliest substances humans transformed by fire into something more valuable. Sumerians and Egyptians initially used lime and gypsum solely for wall and floor plaster. Lime came to play an important role in the manufacture of glass, improving its resistance to weathering, while Egyptians depended on gypsum for the monumental pyramids they built in the third and second millennia B.C.E. Gypsum mortar presented difficulties to masons, however, since it was typically a mixture of burnt (dehydrated) and unburned mineral, which set irregularly. Egyptians used it more as a lubricant for sliding and aligning the great stone blocks of their pyramids than as a cement.

Roman engineers perfected lime mortar at the start of the Common Era, following its early use on Crete and then by the Greeks. Masons heated chalk and seashells (made up primarily of calcium carbonate) to above 900 degrees, at which point they decomposed to lime. The Greeks devised a water-resistant, high-strength cement by mixing lime together with volcanic rocks from the island of Santorini. Later the Romans discovered a particular form of volcanic ash called *pozzulana,* from Mount Vesuvius and the Alban Hills near Rome, that was an excellent source of alumina and silica for hydraulic cement, which had the remarkable and very useful property of setting under water.

Lime cement was first used by Roman engineers as mortar for bricks and stones, and then as the adhesive constituent of concrete, a marvelous pourable stone. Cement reacts with water, in a process called "slaking," in which calcium hydroxide, a hydrated form of lime, is produced with the generation of substantial amounts of heat. "Tobomorite," a gel with an atomic structure between amorphous and crystalline, somewhat like Jell-O, strongly bonds the calcium hydroxide and volcanic ash together.

The bridges and aqueducts that still stand today are silent monuments to the extraordinary skills of Roman engineers two thousand years ago. Ruins of Roman aqueducts abound along the Mediterranean littoral, from Caesarea in today's Israel, to the most spectacular, the Pont du Gard in France, straddling the Gardon River valley to supply Nîmes. Assembled from cut stone held together with pozzulanic mortar, this structure's interior is lined with hydraulic cement. An aqueduct built at Segovia during Trajan's rule at the end of the first century C.E. is still used today to bring water from the Rio Arebeda. Roman expertise is epitomized by the 180-foot-high bridge that spans 115 feet across the Tagus River at Norba Caesarina (renamed Alcantara, meaning "bridge" in Arabic), close to the Portuguese-Spanish border of today. The engineer who erected this bridge for Trajan predicted that it would last forever. His boast was well founded, since today it carries a four-lane highway. Because of the remarkable hardness of Roman mortar, engineers in more modern times speculated that the Romans must have had a secret that was subsequently lost. In the late eighteenth century, French workers made the important but rather prosaic discovery that the excellence of Roman mortar was based on its being thoroughly mixed and then rammed into place, giving it good mechanical contact with the stones or bricks.

In the first century C.E., the Roman writer Vitruvius
gave the earliest explanation of the formation and behavior of cementlike material:

> Stones, as well as all other substances, are com-
> pounded of the elements; those which have most air
> are tender; those which have most water are, by rea-
> son of their humidity, tenacious; those which have
> most fire, brittle. If these stones were only pounded
> into minute pieces, and mixed with sand without
> being burnt, they would not indurate or unite;
> but when they are cast in the furnace, and these
> penetrated by the violent heat of the fire, they lose
> their former solidity; being calcined and deprived of
> their strength they are left exhausted and full of
> pores. . . . [U]pon being replenished with water,
> which repels the fire, they recover their vigor and
> the water entering the vacuities occasions a fermen-
> tation; the substance of the lime is refrigerated and
> the superabundant heat ejected.[23]

Designing monumental buildings for their imperial
patrons, Roman architects learned to overcome the
challenge of spanning huge, ever-increasing areas without using pillars. To accomplish this they utilized barrel
vaults and domes, which are fundamentally arches. Initially they built them with wedge-shaped stones but, as
the areas to be spanned grew larger, the massive wooden
frames needed to support the roof during assembly became impossibly expensive and unwieldy. Confronted
with the increasingly daunting task of erecting larger
roofs, walls, and foundations from stones that had to be

laboriously quarried, shaped, and then transported to the building site, Roman engineers arrived at the ingenious idea of pouring these structures out of *concrete*. A mixture of cement, broken-up rocks, and sand, concrete revolutionized building construction. Sand and rocks were generally available locally and only cement needed to be shipped long distances. Cement bonds the sand and rocks together, producing a synthetic stone, its great advantage being that it could be poured into wooden frames for the desired shape. Using lightweight concrete for vaults and domes in conjunction with a skeleton of arched brick ribs, Roman engineers could now pour a roof in small, independent sections between the ribs.

Roman genius with concrete is epitomized by the Pantheon — the Temple of the Gods — erected by Agrippa in the first century B.C.E., and subsequently burned down and then rebuilt by Hadrian between 115 and 125 C.E. Still standing today in Rome, the Pantheon utilizes brick arches as the framework for a monumental dome, perfectly hemispherical in shape, with an internal diameter of 142 feet, set on twenty-foot-thick cylindrical concrete walls that support the five thousand tons of concrete in the roof. Engineers poured the dome in three horizontal sections, with the concrete in the topmost layer containing pumice stone, the intermediate layer lava rock, and the bottom layer brick fragments. Both the density of the concrete and the thickness of the dome decrease when going from bottom to top, minimizing its overall weight.

Following the fall of Rome in the fifth century C.E., the recipe for high-quality mortar was lost. Saxon and

Norman masons in the Middle Ages formulated poorly mixed mortars, incorporating incompletely burnt lime that had no volcanic ash. Finally, in the fourteenth century C.E., nearly a thousand years after the Romans, the quality of mortar began to improve. Masons rediscovered the Roman composition, which was still considered the best. In the late eighteenth century, the chemistry of cement was investigated systematically by John Smeaton. Following up on his commission to erect a new lighthouse on the Eddystone Rock in England, Smeaton carried out a pioneering study of the best constituents for mortar. His discovery in 1756 that clay must be added to limestone to produce hydraulic cement gave birth to the modern cement industry.

Few innovations occurred in materials used in buildings and bridges from the time of the Roman Empire to the eighteenth century, when Smeaton made his studies and when cast iron was finally produced cheaply enough to use on a large scale. Following the western Roman empire's succumbing to invasions by Germanic tribes during the fifth century of the Common Era, a dark age descended on the portion of the world that we have been exploring. Instead of advances in materials, there were regressions. The quality of glass deteriorated for several hundred years. The mortar used in the Middle Ages was vastly inferior to what the Romans had used a millennium earlier. After the great outpouring of creativity and innovations of the preceding millennia, an outpouring that culminated in the material sophistication of the Roman Empire, came a period of hibernation. Much discussion among historians has taken place about how "dark" the Dark Ages truly were. From a materials point of view, however, there is little question. Experimentation with materials slowed; technology was limited to local concerns and to ensuring immediate survival. Because of the complex nature of the operations required to discover the next generation of materials, a new level of technical expertise was needed. And that expertise would be dependent on the development of the scientific method, or systematic experimentation.

8

Innovations from the East

The nature of papyrus is to be recounted, for on its
use as rolls human civilization depends, at the most
for its life, and certainly for its memory.[24]

— Pliny the Elder, *Natural History*

EUROPE FACED A CHAOTIC political situation following the collapse of the western portion of the Roman
Empire, which had provided substantial stability for
nearly half a millennium. Beginning with the reign of
Augustus, at the end of the first century B.C.E., Europe
was quite stable for two hundred years, the period of the
Pax Romana.

"Stable" is a relative term, of course, since there
were sporadic clashes with the Germanic tribes on the
northern frontier of the empire, as well as two Jewish re-
volts in Palestine, during this period. The first Jewish revolt began in 66 C.E. and ended with the destruction of
Jerusalem and the Temple by Titus, son of the emperor

Vespasian. The isolated desert stronghold of Masada, adjacent to the Dead Sea, fell a few years later. Vespasian proudly recorded his son's triumph by imprinting "Judea Capta" on coins and on the Arch of Titus, still standing today in Rome. The second revolt, led by Bar-Kokhba, lasted from 132 to 135 before being crushed by Roman legions dispatched by Hadrian.

In the third century, the emperor Diocletian, realizing that it was impossible for one person to govern such a huge and diverse domain, split the empire into eastern and western portions. There was increasing turmoil, with the army playing the role of emperor maker, the Visigoths capturing Rome in 410 before moving on to settle in Spain, the Vandals sacking Rome in 455, and finally the fall of the last emperor, Romulus Augustulus, in 476. Lacking the capital to revive exhausted Roman mines, and forced into a barter economy by the scarcity of gold and silver for currency in places such as Britain, large portions of Europe slipped into obscurity. For the next thousand years artisans crafted tools, weapons, and containers from much the same materials as their Roman predecessors. The substances that were discovered during this period, such as gunpowder and paper, which were to have a profound impact on the course of world history, came from the East.

Two of the three major Near Eastern religions emerged during the period stretching from the time of Rome onward for a thousand years: Christianity, by definition, at the start of the Common Era, and Islam in the seventh century. Christianity was born in Palestine during the reign of Tiberius. Christians were sporadically persecuted by the Romans following the crucifixion of Christ in 33.[25] At the beginning of the fourth century, before the battle of Malvian Bridge in Rome, Constantine supposedly saw a flaming cross inscribed with the words "In this sign thou shalt conquer" emblazoned across the sky. Inspired by his vision, Constantine vowed that if he defeated the other competitors for the imperial throne, he would convert the Roman Empire to this new faith. He was victorious and fulfilled his promise.

The rise of Islam in the early part of the seventh century accompanied the transfer of political and military power away from the region once dominated by Rome and back to the Near East, which again came to play an important role in technological development. Advances were now taking place in the Far East, primarily India and China. Following his exile to Medina in 622, Mohammed, the founder of Islam, returned to and occupied Mecca in 630. Following Mohammed's death in 632, Abu Bakr was chosen as caliph and a *jihad,* or holy war, swept out of Mecca and Medina, spreading the faith. Within a century, moving eastward, Islam enveloped parts of the Byzantine Empire and Persia and went all the way to the gates of India; moving westward, it incorporated the remnants of the Roman Empire along the north coast of Africa and in Spain. Included in the Islamic Caliphate now were Alexandria and Antioch, two of the great Byzantine cities. Islamic artisans in these cities absorbed the skills of Christian as well as Persian craftspeople. Trade routes to the Far East passed through the Caliphate, bringing new ideas and exotic wares from India and China. The crafts from this period synthesized Islamic concepts with those from other cultures and were extraordinarily fresh and beautiful. Geometric designs, patterns of plants, and quotations from the Koran were the favored decorations of glassworkers. One Islamic writer wrote of such glass:

> He set before us whatever is sweet in the mouth or fair in the eye. . . . He brought forth a vase, which was as though it had been congealed of air, or condensed of sunbeam motes, or molded of the light of the open plain, or peeled from the white pearl.[26]

A famous product of Islamic culture was the Damascus sword, crafted from steel very high in carbon, 1.5 to 2.0 weight-percent, and containing little sulfur or phosphorus. The steel, imported from India, was called "wootz" steel. These blades had a characteristic etched surface, made by a pattern of light-colored particles of cementite and dark regions of pearlite. To achieve the optimum hardness and flexibility, Damascus smiths devised a unique and particularly grisly heat-treating procedure to control the rate at which the steel is cooled:

> Then let the master workman, having cold-hammered the blade to a smooth and thin edge, thrust it into the fire of the cedarwood charcoal, in and out, while reciting the prayer to the God Balhal, until the steel be of the color of the red of the rising sun when he comes up over the desert toward the East, and then with a quick motion pass the same from the heel thereof to the point, six times

through the most fleshy portion of the slave's back and thighs, when it shall have become the color of the purple of the king. Then, if with one swing and one stroke of the right arm of the master workman it severs the head of the slave from his body, and display not nick nor crack along its edge, and the blade may be bent around the body of a man and break not, it shall be accepted as a perfect weapon, sacred to the service of the God Bal-hal, and the owner thereof may thrust it into a scabbard of asses' skin, brazen with brass, and hung to a girdle of camel's wool dyed in royal purple.[27]

Perhaps the most important concept to pass from East to West in the twelfth and thirteenth centuries was the Arabic system of numeration, which actually was the work of Hindu mathematicians in India. Numbers in Europe had been expressed in Roman notation, which, for large values, was quite awkward. For example, the number 1,372 in Arabic notation is expressed as MCCCLXXII in Roman numerals. Roman notation is particularly unsuitable for simple arithmetic operations such as multiplication and division. Leonardo Fibonacci, an Italian merchant in North Africa, learned of the Indian system of numbers from Arab traders at the end of the twelfth century. In this scheme of numeration, the value of an integer depends on its place in a line of integers. Starting at the far right of the line, the first place is the ones position, moving to the left the second place is the tens position, the third is the hundreds, and so on. Inherent in this notation is the notion that a place position can remain empty. The concept of zero had not been used before in numeration systems. Returning to Italy, Fibonacci wrote a book drawing attention to the Arabic notation, which of course is the system used universally today. Although mathematics is not specifically related to the history of materials, it is difficult to imagine quantitative science developing as it has over the past two hundred years had it employed the clumsy Roman system. In addition to the transmission of Arabic numerals and mathematics from India, other major contributions of the Arab world of that time came through their expertise in alchemy and astronomy.

Recognizing the value of science and technology, the caliphs encouraged their study and commissioned the translation of seminal works from the Greeks and from the Near and Far East. Particularly important for the dissemination of this knowledge was the unifying effect of Islam, with one language, Arabic, dominant from the Iberian peninsula to the Indus River. These Arabic translations were frequently the only source of the great works of the classical scholars available to Europeans.

Seeking to stimulate science and technology, the caliphs established scientific academies, such as the Bayt al-Hikma (House of Wisdom) founded by al-Ma'mun in Baghdad in the early part of the ninth century. Part of the staff's responsibility was to search other countries for valuable manuscripts and translate them. In addition, astronomical observatories in Baghdad and Damascus were attached to the Bayt al-Hikma. In 1004, al-Halim established the Dar al-Kikma (House of

Science) in a Cairo palace. Acting much the same way as modern research universities do today, these academies employed scholars and held seminars to discuss important topics of the day. The mosque was also a center of learning where teachers and students gathered in corners or by pillars. From this came the *madrasa* (school), which evolved into a university, offering instructions in a number of topics. One of the most famous is al-Azhar, founded in Cairo during the tenth century. Frequently housing scholars at their own palaces, caliphs supported them so that they could concentrate on research and writing, very much the way governments today fund basic research at universities and national laboratories.

Europe's emergence from the Dark Ages owed much to knowledge and innovations transmitted from and through Islamic countries. Muslim Spain in particular played a critical role in the technologies that led to the Renaissance. For knowledge to be transmitted, however, a new medium was required.

As we've seen, the first appearance of writing was in Sumer, at the close of the fourth millennium B.C.E. No longer was it necessary to depend on a prodigious memory and an oral tradition to transmit information over distance and time. However, scribes could only record the documents one at a time, and the reproduction of many copies was tedious, time-consuming, and expensive. As a consequence, few people were literate. Creation of new ways to duplicate documents rapidly and inexpensively became as important as the invention of writing itself.

Paper and printing have made profound contributions in the effort both to understand nature's laws and create human laws. The relationship between the written word and the birth, growth, and survival of self-governance cannot be overemphasized. We have only to look to the book burning in Nazi Germany, or, more recently, to the obsessive control over copying machines in Russia and China, to see that rulers of autocratic countries fear written expression. The advent of the printing press and movable type meant knowledge was no longer the exclusive province of a small number of religious men. Books and the information they contained were now accessible to large numbers of people. It is no coincidence that the Protestant Reformation began some fifty years after the printing press first appeared in Johann Gutenberg's shop in Mainz.

Printing was made possible by several advances with quite different kinds of materials. Carving mirror images of the text or figures in wooden blocks, Chinese artisans then coated them with ink (called "India ink" in the West but originally formulated in China, first from pinewood soot and then lampblack), and finally pressed them on paper. Movable type was also fashioned in China when in 1045, early in the Sung dynasty, a commoner named Pi Sheng sculpted individual characters out of clay. In the early 1300s, during the period of Mongol rule, a magistrate named Wang Chen was producing work at such speed that woodblock printing could not keep up, so he designed his own characters and then employed an artisan to carve them from separate wooden blocks. After two years of intensive work, Wang acquired sixty thousand characters, which he used to print one hundred copies of a paper every month. In his classic *Treatise on Agriculture,* printed with woodblocks, Wang included a description of the printing process itself:

> A compositor's form is made of wood, strips of bamboo are used to mark the lines and a block is engraved with characters. The block is then cut into squares with a small fine saw until each character forms a separate piece. These separate characters are finished off with a knife on all four sides, and compared and tested till they are exactly the same height and size. The types are placed in the columns (of the forme) and bamboo strips which have been prepared are pressed in between them. After the types have all been set in the forme, the spaces are filled in with wooden plugs, so that the type is perfectly firm and will not move. When the type is absolutely firm, the ink is smeared on and printing begins.[28]

Printers in China did not reap huge benefits from movable type, however; their language was ideographic, requiring many thousands of characters to express complex thoughts. Since there can be as many as twenty different symbols for each character, it is estimated that a complete set might involve upward of 200,000 Chinese symbols. For an alphabetic language such as ours, fewer than one hundred different symbols are sufficient.

Casting metal in wooden molds, allowing the mass production of individual letter type, was another important innovation. Metal type probably originated in Korea at the start of the fifteenth century, but its use there diminished, again because it was not practical for an ideographic language. By the mid-1400s European printers were casting individual letters with type-metal, a low melting-temperature alloy of antimony, tin, and lead similar to pewter. Pewter had already been used extensively for casting plates and cups, and a great deal of expertise was available to ingenious printers. Individual type was set in even lines and held in place under the pressure of thin wedges, called quoins. A page of these lines, set one above the other, was put into a printing press, whose design was based on a machine that had originally been used to flatten linen.

Printing requires inks that adhere well to paper. Early in the fifteenth century painters in the Low Countries discovered that boiled linseed oil served as a rapid-drying adherent varnish. Gutenberg, frequently credited with being the first Western printer, learned about oil-based ink through his friendships with Flemish artists. Movable type and printing ink were still not sufficient, however, to induce the rapid growth of printing. Before printing could achieve its potential, one more innovation was needed — an easily produced and therefore inexpensive medium on which to print.

Sumerians, Babylonians, and Assyrians employed clay as their writing surface. As early as the third millennium B.C.E., Egyptians took advantage of a common water plant found along the Nile, papyrus, gluing thin strips of the inner bark of the reed lengthwise and crosswise in several layers, then flattening them together, and finally polishing the page to obtain a smooth surface.

Starting in the second century B.C.E., Romans preferred parchment, the dried skins of sheep, goats, and similar animals, with twenty-five sheep required to make a folio, or book, of two hundred pages. Clay was cheap but bulky and not practical for lengthy documents, while papyrus and parchment were expensive to manufacture in large quantities. When scribes wrote on papyrus and parchment, their labor was not the major consideration in the cost of a document. Once paper had made its way from the East, however, scribes became the costly element in the reproduction process, providing a strong motivation to find more economical ways of reproducing written documents. The Greeks used two words for papyrus, *papuros* and *bublos,* the former being the origin for "paper" and the latter, likely derived from the Phoenician port of Byblos, the origin of "Bible."

The earliest remnants of paper come from China, during the Han dynasty in the second century B.C.E. The *Shuo Wen Chieh Tzu* (Analytical Dictionary of Characters), in its definition of *chih,* paper, as "a mat of refuse fibers," suggests why China was the birthplace of paper. The Chinese have a long tradition of pounding and stirring rags in water, using the resulting fibers for quilted clothing. It is likely that rag fibers, accidentally dried on a porous mat, gave the first sheet of paper. The Chinese also have a long history of beating the bark of the mulberry tree into thin sheets for clothing. Large-scale production became feasible when, early in the second century C.E., Chinese artisans realized that fibers from bamboo or hemp served as well as rags for making paper. Papermakers pounded or macerated wood, bamboo, and hemp, weakening the lignin that holds the cellulose fibers together to get clumps of individual fibers. Suspending them in water, papermakers then deposited the fibers onto a screen that held the interwoven mat while the water drained away, leaving a thin, feltlike layer. After drying, the thin mat of fibers was peeled off as a sheet of paper. Smoothing the surface and adding starch and talc to increase its receptiveness to ink completed the process. The paper in this book is made in much the same manner as it was in China more than a thousand years ago.

Knowledge of papermaking was carried westward by trade caravans, as well as by participants in clashes between Muslims and Chinese — Arab tradition has it that the paper industry was founded at Samarkand by a Chinese papermaker captured at the Battle of Talas. Papermaking moved to Baghdad by the close of the eighth century, and by the tenth century paper had completely supplanted parchment in the Islamic world. Baghdad housed more than one hundred bookmakers and thirty-six public libraries by the mid-thirteenth century. Since the earliest Western paper mills were not built until the thirteenth and fourteenth centuries, Baghdad and Damascus were the major suppliers of paper to Europe until the fifteenth century.

Printing was thought to have moved westward as the result of Mongol invasions that swept through Persia, Russia, and then to the borders of Germany in the late thirteenth century, since woodblock printing appeared a short time later in Europe. Card games were in vogue, hence the earliest prints in Europe were wood-block-printed playing cards. The first printing with a press and movable type occurred in 1458 at Gutenberg's shop in Mainz, and the best known of the early printed books was the Bible.

The stage was now set for the next burst of creativity, which would begin in Spain and China and make its way to Western Europe, birthplace of the Industrial Revolution.

9

Stoking the Furnace of Capitalism

> Machines for navigation can be made without row-
> ers so that the largest ships will be moved by a
> single man in charge. . . . Cars can be made so that
> without animals they will move with unbelievable
> rapidity. . . . Flying machines can be constructed so
> that a man sits in the midst of the machine revolving
> some engine by which artificial wings are made to
> beat the air like a flying bird. . . . Machines can be
> made for walking in the sea and rivers, even to the
> bottom without danger.
>
> — Roger Bacon (1214–1294)[29]

OF ALL EUROPE, Spain held to its technological skills
most tenaciously during the period following the fall of
the Roman Empire, perhaps because Roman mining and
smelting expertise had been concentrated there, or perhaps because of the stimulation and stability provided

by the Moors who occupied the Iberian peninsula in the eighth century C.E. Advances in furnace technology in northern Spain led to large increases in iron production toward the end of the first millennium. While the rest of Europe stagnated, Catalonians, first under the Visigoths and then the Moors, developed a new smelter. They built stone furnaces against the side of a hill, filled it with charcoal to the level of the tuyeres, and then stacked iron ore and charcoal in separate columns. A blast of air from bellows through the tuyeres decomposed the charcoal into carbon monoxide, which reduced the ore to metallic iron. Earlier furnaces had yielded fifty pounds of iron in one run, while the Catalan forge could extract 350 pounds. Further increases in capacity by smelters in Austria and Germany led to the development of the *Stückofen* or "wolf" furnace, which by the fifteenth century was producing as much as 700 pounds of iron in a run. A pair of hand bellows provided the blast of air to the Catalan forge, but human-power no longer sufficed for the *Stückofen*. A new source of power was needed to generate blasts of air that could penetrate to the tops of these tall furnaces, which reached heights of ten or more feet.

The size of the *Stückofen* and the quantities of ore needed to feed it brought demands for mechanical devices to crush and hoist ore, forge hot iron, drain mines, and blow air. By the eleventh century, waterpower, which had been used for grinding corn since the time of Rome, was also being widely applied for hoisting and crushing ore, and for blowing air during smelting. Waterpower enabled enormous increases in the amount of iron produced in the *Stückofen,* which in turn contributed to the maintenance of stable iron prices throughout the Middle Ages. By the eleventh century, wind power was also being exploited. One of the keys to Europe's emergence from the Dark Ages was large-scale mechanization based on both waterpower and wind power.

Monasteries, such as those built by the Cistercians, were extensively involved in constructing waterwheels. Seeking to mechanize their activities as much as possible — not wanting to waste time that could be better spent in prayer — Cistercians had both the resources and motivation to exploit waterpower. An early description of their abbey at Clairvaux gives a good picture of the success of their endeavor:

The river . . . gushes first into the corn mill where it is very actively employed in grinding the grain under the weight of the wheels and in shaking the fine sieve which separates flour from bran. Thence it flows into the next building, and fills the boiler in which it is heated to prepare beer for the monk's drinking, should the vine's fruitfulness not reward the vintner's labor. But the river has not yet finished its work, for it is now drawn into the fulling-machines following the corn mill. In the mill it has prepared the brothers' food and its duty is now to serve in making their clothing. This the river does not withhold, nor does it refuse any task asked of it. Thus it raises and lowers alternately the heavy hammers and mallets, or to be more exact, the wooden feet of the fulling-machines. When by swirling at great speed it has made all these wheels revolve swiftly it issues foaming and looking as if it had ground itself. Now the river enters the tannery where it devotes much care and labor to preparing the necessary materials for the monks' footwear; then it divides into many small branches and, in its busy course, passes through the various departments, seeking everywhere for those who require its services for any purpose whatsoever, whether for cooking, rotating, crushing, watering, washing, or grinding, always offering its help and never refusing. At last, to earn full thanks and to leave nothing undone, it carries away the refuse and leaves all clean.[30]

Cistercians also pioneered the smelting and forging of iron with the help of waterpower.

As furnaces grew taller and the blasts of air became more powerful, ore was exposed to charcoal at higher temperatures and for longer periods of time. While trying to smelt ore to obtain wrought iron, smelters began to notice liquid iron dribbling out of their furnaces. Serendipitously they had produced a liquid containing 3 to 4 weight-percent carbon, which melts at the relatively low temperature of 1130 degrees and could be cast in molds, hence the name cast iron. As early as the first millennium B.C.E., ironworkers in the Near East were familiar with cast iron, but discarded it because it was too brittle for blacksmiths to forge. Before the Common Era, only artisans in China exploited cast iron, principally to make bells and ceremonial objects such as mirrors, incense burners, and food and drink containers.

Stephen L. Sass

In the tenth century, when wrought iron (containing less than 0.1 weight-percent carbon) was in great demand, smelters in Europe dumped the cast iron back into their furnaces. Since iron with small amounts of carbon melts in the vicinity of 1500 degrees — a temperature not obtainable in the early centuries of the second millennium — smiths had no alternative but to form spongy masses of wrought iron by hot forging, or by hammering at high temperatures. However, smelters came to appreciate that the previously undesirable cast iron could be used profitably in church bells, for which there was an increasing demand due to extensive cathedral building, and later, beginning in the fourteenth century, for cannons and cannonballs. A new market emerged, and the *Flussofen,* or "flow oven," was devised in the early part of the fourteenth century in northern Europe and France. In the *Flussofen,* ore and charcoal added at the top slowly sank to the bottom, reacting to produce iron, which absorbed large quantities of carbon and finally emerged as cast iron. All this was done in a continuous process. Thus was born the ancestor of the modern blast furnace.

To assemble the large quantities of machinery they needed, mine owners and ironmasters required substantial investments of capital. During Roman times, mines, such as those at Rio Tinto in Spain, were owned by the state, which could afford to mechanize them. The situation in Europe was quite different in the fifteenth and sixteenth centuries, when the princes and kings who owned mines frequently had little money to support their operations. New institutions and new methods

151

arose to finance these large-scale operations — and with them came the birth of capitalism.

Leading the way in the financing of large-scale mining projects in central Europe were the Fuggers, Welsers, and other banking families of Augsburg and Nuremberg in southern Germany. The House of Fugger ran wholly self-sufficient mining and smelting operations. Enormous metallurgical works called *Saigerhütten,* comprised of buildings more than 100 yards long, with a series of rooms for furnaces, bellows, forges, and hammers, were frequently powered by water. In these works, silver was extracted from copper ores shipped hundreds of miles from the Fugger mines in Tyrol and Hungary, using the old Roman process of liquation. To ensure the efficient operation of their vast works, the visionary Fuggers trained their mine and smelting foremen in a *Bergschule,* a "mountain school." The enormous influence and arrogance of these merchants is best epitomized by the epitaph dictated by Jacob Fugger:

> To God, All Powerful and Good! Jacob Fugger of Augsburg, ornament to his class and to his country, Imperial Councilor under Maximilian I and Charles V, second to none in the acquisition of extraordinary wealth, in liberality, in purity of life, and in greatness of soul, as he was comparable to none in life, so after death is not to be numbered among the mortal.[31]

In England the Crown controlled all precious metal mines, but it too had insufficient means for their mechanization. Hochstetter, a German engineer brought over to run copper mines in Cumberland, received a charter from Queen Elizabeth I in 1565 in recognition of his pioneering achievements. Hochstetter then "determined to join with him in company divers others, and in that respect doth mean to make divident of the commodities and profits." Based on this charter, which was a grant of incorporation to the shareholders, Hochstetter gained financial backing for the Company of Mines Royal from capitalists eager to invest in mines with a potential for producing precious metals. He also astutely gave fifty shares to a group of powerful people, including the earls of Pembroke and Leicester. His enterprise illustrates the beginnings of joint-stock companies: a company owned by, yet separate from, the shareholders, whose ownership came in the form of transferable shares. In the early days of mining and smelting, owners either worked or at least supervised the running of their mines. Under this new financial arrangement, shareholders were often absentee partners. They received "divident [sic] of the commodities and profits," in Hochstetter's words, either in raw ore or, if it had been smelted and sold as metal, cash.

Forced to borrow heavily from wealthy merchants to operate their mines, the kings of Europe nevertheless did not lose control of them. In fact, monarchs frequently confiscated properties owned or leased by bankers. During the twelfth century, Henry II of England was kept solvent by Jewish financiers, including Aaron of Lincoln. However, in the thirteenth century, so impoverished were the Jews of England by heavy taxes and fines that Edward I, deciding they had been milked dry, expelled them in 1290. Italian bankers suffered a similar fate in England during the fourteenth century. Conditions across the Channel were no better. Charles VII of France, heavily in debt to the famous merchant Jacques Coeur, arrested him and confiscated three of his lead, copper, and silver mines in 1453.

The ability of financial institutions, such as banks and private companies, to amass substantial amounts of money was crucial to the development of large-scale, machine-intensive industries. As the need for such investment grew, merchants and banks in England were willing and able to provide these funds, more so than in other countries.

Up to the 1500s, mining and metallurgical industries were generally more advanced on the continent than in Britain. Silver was the most important precious metal then, and though Germany and France had plentiful ore reserves, Britain had few; consequently, the Crown never viewed mines as great sources of wealth. As the value of the silver mines on the Continent grew, so too did the power of princes and kings, and their increasingly restrictive laws made it more and more difficult for private investors to profit from mining investments, as Jacques Coeur learned.

Henry VIII would certainly have enjoyed being as absolute a ruler as the French king, but by this time he was constrained by peculiarly British traditions and institutions. The Magna Carta, signed by King John in the early part of the thirteenth century, gave birth to a Parliament that over six centuries would transform an absolute monarchy into a constitutional one. While recognizing Henry as king, Parliament steadfastly defended the right of common people to own property and served as a restraint on his behavior. In addition, while the Crown owned the few precious-metal mines in Britain, others that produced less valuable substances were in private hands. When British royalty sought new sources of income, they ignored the iron and coal mines, which were then free to be developed by commoners without fear of confiscation. By the time coal and iron had become the key materials of the Industrial Revolution, restraints on the British sovereigns were strong enough to keep their acquisitive fingers off the now valuable mines.

Having played little or no part in the innovations that were so critical to the advance of technology up to this time, western Europe assumed a central role in the middle of the second millennium of the Common Era. The conjunction of a variety of innovations and circumstances explain why the Industrial Revolution occurred where it did — the development of novel means of financing large capital outlays was one reason and the special nature of the monarchy another. As we shall see shortly, other factors, such as the presence of iron and coal supplies, were even more important to large-scale industrialization.

By the sixteenth century demand for iron had increased dramatically, driven in part by the appetite of the military for new weapons, primarily firearms and cannons. The inventions of gunpowder and cannon in China were among the most remarkable developments of the period. Experimenting with mixtures of chemicals that could burn, alchemists in China warned against dangerous formulas in the *Chen Yuan Miao Tao Yao Lueh* (Classified Essentials of the Mysterious Tao of the True Origin of Things) from the ninth century:

> Some have heated together sulfur, realgar [arsenic disulfide], and saltpeter with honey; smoke [and flames] result, so that their hands and faces have been burnt, and even the whole house burned down.[32]

These ingredients are the basic components of gunpowder, which is a mixture of saltpeter (potassium nitrate), sulfur, and charcoal.

Byzantium resisted the onslaughts first of Islam and then of the Crusaders for as long as it did because of "Greek fire," a form of flamethrower that pumped out volatile liquids in bursts of flame. Gunpowder probably originated with similar incendiary weapons based on the use of naphtha, which is volatile. As early as 919, during the chaotic transition between the Tang and Sung dynasties, the Chinese also used such pumps, but added a slow gunpowder fuse to ignite the liquid. By the next century they had learned how to put gunpowder and large amounts of saltpeter in casings of bamboo to make "thunderclap bombs," meant as much to scare as to harm the enemy. Later they used iron, which shattered and killed men and horses. By trial and error the Chinese discovered that the optimum composition for gunpowder was 75 weight-percent saltpeter, the remainder being made up of approximately equal quantities of sulfur and charcoal.

Packing gunpowder into hollow bamboo stems

open at one end and then attaching them to spears, soldiers fashioned fire lances, a rudimentary form of flamethrower. Constricting the open end of the bamboo and turning the tube around, they then invented the rocket. First devised toward the end of the twelfth century, at the high point of Sung rule, the early rockets probably terrified rather than killed people. The key advance in all these innovations, though, was the use of cylinders to hold the gunpowder. Continuing their improvements in the twelfth century, the Chinese replaced the bamboo in the fire lance, first with bronze and then with cast-iron tubes. Finally, adding stones to shoot out of the tube, the Chinese invented the forerunner of the cannon.

The first projectiles, small stones loosely packed above the gunpowder, evolved into balls that fit snugly inside the tube. The carbon and sulfur in gunpowder are excellent reducing agents, meaning that they react strongly with oxygen, forming stable molecules while releasing large quantities of energy. Potassium nitrate has three oxygen atoms out of a total of five atoms in the compound, so that it provides plenty of oxygen for combustion. The reaction of sulfur and charcoal with large amounts of potassium nitrate is explosive, generating a tremendous quantity of heat. Upon ignition, solid gunpowder transforms completely to gas at temperatures well above 3000 degrees, with an accompanying volume increase of more than three thousand times. The gas volume increase shot the ball out of the tube at high velocity — and the world had its earliest cannon.

The first cannon appeared in China toward the end

of the thirteenth century, and showed up in the West a few decades later. One of the earliest uses of gunpowder in Europe was at the Battle of Crécy in 1346, where the English, outnumbered more than three to one by the French, defeated them allegedly using cannon in combination with the longbow (which actually played the more crucial role). This decisive battle, which started England on its rise to dominance in world affairs, also signaled the decline in importance of armored knights as the ability to kill at long range was refined. China clearly knew of gunpowder before the West, since the Chinese used it as a fuse as early as 919 and its preparation is described in the *Wu Ching Tsung Yao* (Collection of the Most Important Military Techniques), dating from 1044:

> 1 chin [a pound, roughly], 14 ounces of sulfur, 2½ chin of saltpeter, 5 ounces of charcoal, 2½ ounces of pitch, and the same quantity of dried varnish are powdered and mixed. Then 2 ounces of dry vegetable matter, 5 ounces of tung oil, and 2½ ounces of wax are stirred into a paste, and finally all the ingredients are mixed together.[33]

Because this concoction contained large quantities of petroleum products, it was probably meant more for fireworks than for explosives.

By adding poisonous chemicals and powdered lime to their bombs, the Chinese produced truly vicious weapons, foreshadowing chemical warfare, which was

used by the Sung general Yo Fei when he was attempting
to take the bandit Yang Tao by surprise in 1135. He relates:

> The army also made "lime-bombs." Very thin and
> brittle earthenware containers were filled with poi-
> sonous chemicals, lime, and iron calthrops. In com-
> bat they were used to assail the enemy's ships. The
> lime formed clouds of fog in the air, so that the rebel
> soldiers could not open their eyes.[34]

Knowledge of gunpowder likely traveled westward
along the Silk Road — caravan trade routes linking
Byzantium and the Islamic Caliphate to the Far East. It
ran from China to Samarkand, then skirted the southern
tip of the Caspian Sea, and thence on to Baghdad and
Antioch. Islamic generals resisting the Crusaders utilized
this technology to excellent effect. Probably the key factor in the Muslim victory at the Battle of al-Mansura in
1249, during which Louis IX of France was captured,
was incendiary weapons. Shortly before cannons appeared, Muslim fighters found ingenious methods to
torment Crusaders, including hurling explosive firepots
into their encampments with giant crossbows. A French
officer named De Joinville vividly recalled the terrifying
night attacks:

> It was like a big cask and had a tail the length of a
> large spear: the noise it made resembled thunder
> and it appeared like a great fiery dragon flying

through the air, giving such a light that we could see in our camp as clearly as in broad daylight.[35]

Following the fall of the Christian fortress at Acre to the flaming onslaughts of Islamic military engineers in 1291, the ill-fated Crusades finally drew to a close. It was around that time that Islamic armies started using cannons, and Europeans likely learned of their power during clashes with Arab forces in Spain. Monstrous bronze cannons with bores of three feet and balls weighing up to almost 600 pounds contributed to the capture of Constantinople by the Ottomans in 1453 and the collapse of the Byzantine Empire.

Cannons in both the East and the West were first cast from bronze because it melted at the low temperature of 1000 degrees. Metalworkers also fabricated cannons by welding together wrought-iron bars along the axis of the bore and then shrinking metal hoops around the bars for additional strength. These models were prone to explode if the gunpowder charge was too large, as in the unfortunate blast that killed James II of Scotland in 1460. The Ottoman Turks employed cannons so massive that they frequently had to be cast *in situ*, since they were too large to be moved. A particularly monstrous bombard designed to batter the walls of Constantinople in 1453 cracked after only one firing. Sitting within the walls of the Kremlin is a monumental cannon that has never been fired.

England began casting iron cannons in the mid-sixteenth century. For several hundred years their manufacture was restricted to the Weald of Sussex, and so successful were the English that cannons soon became important for trade, both legal and illegal. And, in fact, the Privy Council was besieged with complaints, such as this one dating from 1573:

> [M]erchante shippes . . . doe finde themselves mar-
> vailouslie molested and otherwhiles robbed by rea-
> son of the great store of ordenance that hath both
> been convayed and solde to strangers out of this
> Realme, whereby their shippes are so well ap-
> pointed that no poore merchante shippe maie pass,
> thorow the seas.[36]

Despite laws passed by Elizabeth I limiting the export of cannons, so lucrative was their trade that it was impossible to suppress smugglers. When the Spanish Armada weighed anchor to invade Britain in 1588, most of its 2,400 guns were returning to their place of manufacture.

As the demand for weapons increased, consumption of iron grew dramatically. On a daily basis, the French used more than 12,000 iron cannonballs during the siege of Magdeburg in 1631. Because of the increased costs of both charcoal and labor, the price of iron became inflated. Ironmasters depended completely upon wood for their charcoal, while shipwrights consumed vast quantities of timber for merchant vessels plying the far-flung trade routes connecting Europe with the New World and the Far East, and for the ships of the line of the European powers pursuing markets and colonies. (Spain's economy was fueled by the vast amounts of gold and silver flowing from its colonies in the Americas. Following Pizarro's conquests, rich silver mines were discovered at Potosí in present-day Bolivia, which produced 60 percent of the wealth sent to Spain in the sixteenth and seventeenth centuries. As a consequence of this flood of precious metals, the price of silver fell, and toward the end of the sixteenth century many European mines closed.) This was in addition to wood consumed as fuel for cooking, baking, brewing, and heating. Severe deforestation in the south of England led to the decline of iron making there, and smelters shifted away from the Forest of Dean to fresh sources in Wales and farther north in the Midlands.

From the standpoint of volume, wood was the most widely used material in Europe, both for energy and construction. This was true through much of early human history except in Mesopotamia, where it was scarce, and in Rome, where major buildings and aqueducts were made of stone and concrete. Certainly up to the time we are discussing, wood had no peer in the building of ships and early machines, and as a source of energy. In addition, wood ash was an important source of potash used in glassmaking. We'll look at wood later, though briefly, because humans transformed only its shape, never its properties.

In the mid-sixteenth century Queen Elizabeth passed a series of laws limiting the use of wood for smelting. Between 1588, the year of the Spanish Armada, and 1630, the price of wood in England tripled. Charcoal increased from being one-half to more than three-quarters the cost of smelting iron. During much of the seventeenth century, England accounted for half of the 60,000 tons of iron produced annually in Europe, but by the start of the eighteenth century, as the timber famine worsened, the number of English forges decreased by more than 90 percent. It was becoming quite apparent that the survival of the iron industry (and the entire process of industrialization) depended on finding a cheap substitute for wood, both as a fuel and a source of carbon for smelting. Industrialized nations are in much the same situation today, given that the world's supply of oil is recognized as finite and that a new convenient and economical source of energy will need to be found sometime during the twenty-first century.

By the seventeenth century, Europe and England in particular were beginning to burn coal, a mineral consisting of carbon, organic resins, and traces of elements from decomposed plants. However, early attempts to smelt with coal failed because it contained so many impurities, such as sulfur, that were absorbed by the iron. Pure iron, or iron containing a small amount of carbon, is quite ductile, meaning that it undergoes a large amount of plastic deformation before breaking, which is why blacksmiths preferred working with wrought iron. Unfortunately, when small quantities of sulfur are present in iron, the sulfur atoms are strongly attracted to grain boundaries. As little as one-quarter of an atomic layer of sulfur at grain boundaries — when one out of four atoms in the plane of the interface is sulfur — makes iron so brittle that it fractures or breaks under small loads, such as during forging. Like glass, brittle iron fails catastrophically.

In addition to smelters, brewers and bakers also

continued to search for new sources of fuel to replace wood. By the mid-seventeenth century, brewmasters, failing in their early attempts to use coal for drying malt — beer absorbed evaporating gases and acquired an unpleasant taste — learned to drive off coal's volatile components, such as coal tar and resins, as well as the sulfur, leaving behind coke, a hard, porous gray residue consisting of more than 80 percent carbon. Coking coal was similar to the process of making charcoal from wood. Setting a pile of coal on fire, coke makers then covered it to reduce the amount of oxygen present.

Fifty years passed before Abraham Darby success-fully employed coke to produce cast iron in 1709. Leas-ing ironworks at Coalbrookdale near the Severn River, Darby was fortunate that the coal mined close to his foundry contained very little sulfur. The cost of charcoal had been more than three-quarters of the cost of smelt-ing, as noted earlier. Now, so cheap was coke that Darby could sell cast-iron pots and kettles at prices low enough to make them accessible to common folk. By so doing he opened up a consumer-driven economy. It took another forty years before coke was widely accepted as a substi-tute for charcoal, since most coal was contaminated with much more sulfur than that used by Darby. The Darby family was remarkably innovative. Three gen-erations made their mark on the iron industry. In the 1750s, Darby's son, Abraham II, succeeded in expand-ing his father's business into the manufacture of nails, a difficult market to penetrate because carpenters stub-bornly preferred nails fashioned from charcoal-smelted

Calcite statue of Gudea, early ruler
of the ancient city of Lagash at the
end of the third millennium B.C.E.

Clay cuneiform
tablet from the early
third millennium B.C.E.

Copper and copper-arsenic artifacts from the Judaean Desert Treasure, dating from the fourth millennium B.C.E. (COURTESY ISRAEL MUSEUM, JERUSALEM)

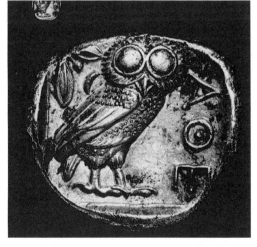

Silver tetradrachma, showing the owl of Athens, struck ca. 440 B.C.E. This was the basic currency unit of the ancient Greek economy. (COURTESY SPINK & SON, LTD., LONDON)

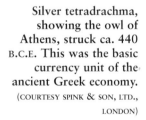

Roman blown-glass bottles from the first century C.E. (COURTESY TOLEDO MUSEUM OF ART)

Schematic drawing of a *dislocation*, showing the extra half plane of atoms. This "defect" is what makes metal crystals weak; if many are present, the crystals become strong again.

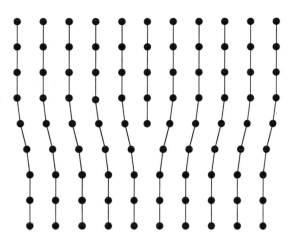

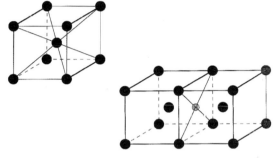

Body-centered cubic unit cell of pure iron (far left); two unit cells showing the location of a carbon atom within an octahedral site (left). The distortions around the carbon atom are the origin of steel's great strength.

Roman aqueduct in Segovia, Spain, from the first century C.E.

Dome of the Pantheon in Rome, built by the emperor Agrippa
and rebuilt by Hadrian in 115–125 C.E.
(COURTESY JOHN ROSS/PHOTO RESEARCHERS)

"Greek fire" being used in a naval battle, as depicted in a tenth-century
Byzantine manuscript. (COURTESY OXFORD UNIVERSITY PRESS)

The early printing process in China, ca. 1300.
(COURTESY CAMBRIDGE UNIVERSITY PRESS)

Iron bridge across the Severn River in England, built by
Abraham Darby III in 1779. Today it is used as a footbridge.
(COURTESY R. PARKER/VISION INTERNATIONAL)

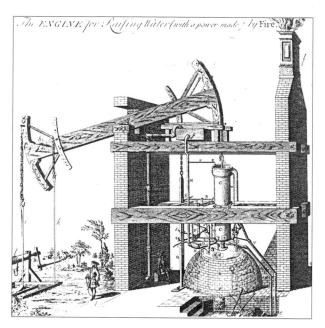

The ENGINE for Raifing Water (with a power made) by Fire.

Earliest known illustration of Thomas Newcomen's 1712 steam engine. (COURTESY LORD BRIGGS, FORMER PROVOST OF WORCESTER COLLEGE, OXFORD)

Print from Agricola's *De Re Metallica*, showing a sixteenth-century mine. (COURTESY DOVER PUBLICATIONS)

Platinum mask from Ecuador
dating from the ninth century.
(COURTESY MUSEUM FÜR
VOLKSKUNDE, STAATLICHE MUSEUM
PREUSSISCHER KULTURBESITZ,
BERLIN)

Charles Goodyear (1800–1860),
inventor of vulcanization, which
made rubber usable.
(COURTESY PICTURE COLLECTION, THE BRANCH
LIBRARIES, NEW YORK PUBLIC LIBRARIES)

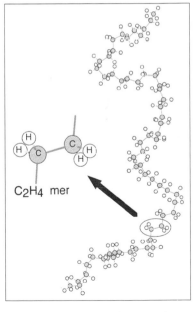

C_2H_4 mer

Schematic drawings showing the
polyethylene polymer chain and its
basic building block. (COURTESY
C.K. OBER, CORNELL UNIVERSITY)

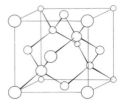

Atomic structure of diamond, showing the tetrahedral bonding that gives diamond its remarkable properties. (COURTESY CLARENDON PRESS, OXFORD UNIVERSITY PRESS)

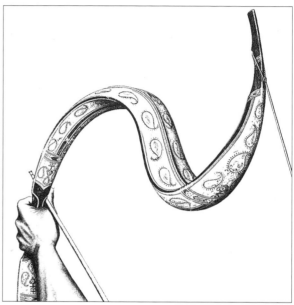

Composite bow, made up of sinew, wood, and horn, first fashioned in the third millennium B.C.E. (COURTESY HANK IKEN)

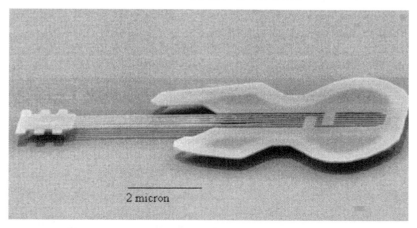

2 micron

Silicon guitar, about 1/25th the diameter of a human hair. Although a dazzling demonstration of the current ability to miniaturize, it is not—alas—playable. (COURTESY D. W. CARR AND H. G. CRAIGHEAD, CORNELL UNIVERSITY)

iron. And in 1779 *his* son, Abraham III, spanned the Severn River with the first iron bridge by casting seventy-foot-long ribs, demonstrating that iron could be used for large structures. The bridge still stands, a monument to Darby's ingenuity, in a place called, appropriately enough, Ironbridge. The descendants of Abraham Darby continued to operate his Coalbrookdale foundry until well into the twentieth century.

By the end of the eighteenth century, charcoal furnaces had nearly disappeared in Great Britain. Castings of iron replaced comparable implements fashioned from brass, wrought iron, and wood, because they were both cheaper and of higher quality. Cast iron completely supplanted wrought iron, since casting saves both the labor and fuel costs of forging. By the end of the nineteenth century, smelting of iron with coke had grown into a worldwide enterprise, with an output approaching 100 million tons annually.

Coal now became the primary energy source of Europe. It is impossible to imagine the Industrial Revolution, based as it was on machine-powered industries for the mass production of a variety of consumer goods, occurring without coal. Britain's central role in developing such industries was largely dependent on its ample supply of coal. (As early as the tenth or twelfth centuries, the Chinese had scaled up their production of iron based on blast furnaces, using coal for both coke and fuel. By the time Marco Polo visited China in the thirteenth century and sent reports back to the West of the burning of "black stone," iron production there had already dwindled. Despite their earlier use of both cast iron and coal, the Chinese did not capitalize on them nearly as much as the West.) For two hundred years Britain was the world's leading coal producer, until finally being surpassed by the United States at the close of the nineteenth century.

In the centuries prior to the Industrial Revolution, a host of problems had to be addressed before coal could be available in large quantities. After the more accessible surface outcroppings were exhausted, mine shafts had to be sunk deeper and deeper. Flooding became an overriding concern. Roman mines at Rio Tinto had been drained by a series of water wheels powered by horses or slaves, but this was no longer feasible in the eighteenth century. Wind power was not dependable except in the Low Countries, where it was used extensively for pumping water and grinding flour. Waterpower was not reliable year-round and also was not always available where it was needed, since waterwheels are generally built in narrow, easily dammed valleys to obtain a sufficient height of water.

Colliery owners sought new sources of power to pump out their mines, and attention turned to the energy available in a jet of steam. Everyone knew that boiling water in a kettle produces steam, which whistles out of the spout and causes the lid to rise and jiggle, so here was a physical phenomenon that had the potential to lift heavy loads. The question was how do so economically. Hot air had been put to work as early as the first century of the Common Era when, in a clever arrangement described by Hero of Alexandria, the heat of a sacrificial fire caused air in a hollow altar to expand and open temple doors automatically. The effort to build machines to drain mines culminated in the development of steam engines, which eventually replaced falling water as the primary source of power behind the Industrial Revolution.

A milestone on the path to steam power was the discovery that the air we breathe exists under pressure. Galileo and his student Torricelli demonstrated that the earth's atmosphere exerts a pressure equal to the weight of a vertical column of water approximately thirty feet high. (Atmospheric pressure is 14.7 pounds on a square inch, which means that if a hollow cube whose edges one inch in length were evacuated so that it contained a vacuum, each face would bear a load of 14.7 pounds.) An early demonstration of the atmosphere's pressure was the famous experiment by Otto von Guericke, mayor of Magdeburg in Prussia. Von Guericke joined together two hemispheres to make a sphere that was subsequently evacuated of air, and hence held together only by atmospheric pressure. These hemispheres could not be pulled apart by two teams of eight horses. But how could this physical phenomenon be made useful? In 1690 Denis Papin, a French Huguenot, suggested a simple device based on the use of both atmospheric pressure and steam. Originally trained to be a medical doctor, Papin had worked with the Dutch astronomer Christian Huygens, who inspired him to look for new sources of power. To avoid persecution for his Huguenot beliefs, Papin eventually escaped to England, bringing along his interest in discovering how to use atmospheric pressure to do work.

Papin's engine consisted of a vertical cylindrical tube closed at its bottom end and containing a piston. When a small quantity of water was introduced under the piston, which was resting just above the bottom of the tube, and heat applied to evaporate the water and form steam, the piston moved up toward the top of the cylinder, where it was held by a catch. The cylinder was then cooled and the steam inside condensed back to liquid water, creating a vacuum under the piston. After the catch was released, the pressure of the atmosphere above the sliding piston forced it down, because the pressure inside the cylinder had dropped. This downward movement, resulting from the difference in pressure between the atmosphere and the vacuum, could be used to lift a heavy weight. Papin never built a practical version of his engine, but he was the first to point out how steam could be used to move a piston and, therefore, the first to establish the fundamental principle underlying future steam engines.

Papin's engine illustrates how to calculate the work done by such a device. The downward force exerted by the piston can be used to lift a weight by means of a simple pulley. Because there is a vacuum inside the cylinder, the force exerted on one square inch of the piston by atmospheric pressure is 14.7 pounds and the downward force is determined by multiplying the area of the piston head, measured in square inches, by 14.7 pounds per square inch. If the area of the piston head is 113 square inches, corresponding to a circle twelve inches in diameter, then the weight that can be lifted is 1,660 pounds. But merely supporting a weight by means of a pulley isn't doing work. A table could just as well support the weight. Work gets done when the weight moves a distance. So in the case of Papin's engine, to determine the amount of work being done, the downward force exerted must be multiplied by the distance that the weight is lifted (and the piston moves). If the piston moved two feet, then the work is two feet times 1,660 pounds, or 3,320 foot-pounds. Because the work done, or equivalently the energy expended, has units of force multiplied by distance, ingenious people devised schemes to lift a heavy weight while exerting a small force by having it act over a long distance. This, of course, is the principle of the lever.

In 1698 Thomas Savery, a prolific inventor born into an affluent Devonshire family, built a water-lifting apparatus that took advantage of the vacuum formed by the condensation of steam. Since Savery's approach could only raise water twenty-five to thirty feet, he combined the condensation of steam with steam under pressure to achieve approximately double this height. A series of his devices, stacked one above the other, was used to lift water in tall buildings. Most mines were two hundred to three hundred feet deep, however, and placing boilers every fifty feet underground was impractical, so Savery's engine was never used successfully where it was most needed.

In 1712, an ironmaster who also came from Devonshire, Thomas Newcomen, succeeded in building the first practical steam engine based on the principle Papin had enunciated. Newcomen injected steam from a boiler into the cylinder, causing the piston to rise, and then sprayed a jet of water inside the cylinder to cool the steam, condensing it back to water. His engines were first fabricated from brass, an alloy of copper and zinc used by the Romans at the start of the Common Era. Newcomen's engines were marketed to miners for pumping water out of shafts and for raising water for use in water wheels, since the piston in his steam engine only produced up-and-down movement transmitted to the water pump by a rocking beam, not the rotary motion needed to power flour mills, for example. Until the cylinder and piston could be made from cast iron, whose cost was one-third that of brass, Newcomen's engines were not widely used. For many years Abraham Darby's foundry was the only shop in England that could fabricate these parts from iron.

Newcomen's engine was inherently wasteful of energy, because the hot steam was cooled and condensed in the same cylinder into which it had been injected to raise the piston. In the following cycle, before the piston could rise, the cold cylinder had to be reheated again by the next injection of steam. Thus, only part of the steam was available to move the piston and do useful work, while the rest was used to reheat the cylinder. This occurred when the steam condensed to liquid water on contact with the cold cylinder walls and gave up its latent heat of condensation, the amount of energy needed to be removed from gaseous H_2O molecules so that they can condense and form liquid water. The portion of the steam condensed does not contribute to moving the piston and is wasted. Newcomen's engine used less than 1 percent of the energy available from burning coal to do useful work, wasting more than

99 percent. Since coal was plentiful at collieries, Newcomen's engines were extensively employed to drain the shafts. Tin mines in Cornwall, on the other hand, could not afford the engines, because buying and shipping coal was too expensive.

After examining a Newcomen engine he had been given to repair, James Watt, an instrument maker working at Glasgow University, took the next important step toward an economical steam engine. Realizing that the engine could be made much more efficient if the cylinder containing the piston were kept hot, Watt's solution was to connect a separate steam condenser to the main cylinder by means of a valve and piping. He also came up with the clever idea of using low-pressure steam instead of the atmosphere to push the piston down. Watt was granted a patent for "A New Method of Lessening the Consumption of Steam and Fuel in Fire Engines" in 1769. His first partner, John Roebuck, acquired two-thirds interest in this patent. After Roebuck went bankrupt, Watt sought another business partner and met Matthew Boulton, a manufacturer from Birmingham, who took over Roebuck's share, which the bankruptcy receivers valued at one farthing. Boulton provided both financial backing and business acumen, and Watt began manufacturing engines in the 1770s. At least three times as efficient as Newcomen's, his engine reduced coal consumption by more than two-thirds. In what turned out to be a wise financial arrangement, a mine owner purchasing an engine from Boulton and Watt also had to pay them one-third of the savings in fuel costs over other engines.

With the exception of Newcomen's early brass models, all other engines were cast from iron. No other material that could be made into cylinders able to hold both hot steam and a vacuum as inexpensively was available. All the connecting pipes had to be vacuum-tight and therefore also fashioned out of metal. To maintain the vacuum, the piston needed to fit snugly inside the cylinder, requiring accurate boring. Fortuitously, one of Watt's first engines was installed to blow air into blast furnaces at the ironworks of John Wilkinson, who had obtained a patent for his cannon-boring mill in 1774. His mill was exactly what was needed to fabricate engine cylinders. As Watt's steam engine became more popular, a need developed to convert the up-and-down movement of the piston into a rotary motion. Watt invented the sun-and-planet arrangement of gears to do this. Earlier gears for water wheels fit together loosely and were cut from wood. Watt's gears needed to be machined to close tolerances, and only metals could satisfy this requirement. Clockmakers such as John Harrison, the inventor of the chronometer, had known this for a long time.

All this illustrates how inventions emerge from a dynamic interplay between the needs of the times and the creativity of craftspeople. Here, for instance, are the key factors motivating Watt's historic achievement, as well as their significant consequences:

- To produce large quantities of iron (as well as bread, beer, salt, and so forth) Europe needed to find a new source of carbon and energy to substitute for wood. Coal fit the bill.

• To produce large quantities of coal, entrepreneurs sank deep mine shafts, but flooding necessitated the development of more efficient means of pumping water. A new source of power was needed and eventually the iron steam engine was invented.

• To bring coal and iron ore together, businessmen needed inexpensive transportation, as mines and smelters were generally in different locations. Thus there was pressure to build a network of canals for cheap transport.

• To deliver coal to central shipping points, such as the harbor at Newcastle-on-Tyne, mine owners began using horse-drawn carts that ran on wooden rails. In time, steam engines replaced horses as motive power for these carts and the railroad was born in the nineteenth century. This in turn generated an enormous demand for inexpensive, high-strength rails, satisfied first by iron and then by steel.

Beginning with a timber famine and iron smelting, and ending with a huge and ever-increasing demand for iron, a cycle of need and innovation developed. That cycle continues today.

One other aspect of the steam engine's history should draw our attention. Before devising his conceptual model, Papin worked with Huygens on the idea of using an explosion of gunpowder to heat gases and move a piston; when the gas cooled, a vacuum was created. Atmospheric pressure would then push the piston back down. Papin came to realize that the gases remaining in the cylinder after the explosion made it impossible to attain the desired vacuum, and so he abandoned this approach and turned, instead, to steam to move the piston. An explosion beneath a piston within a cylinder was the basis of the internal combustion engine invented two hundred years later, at the close of the nineteenth century. A small quantity of gasoline or diesel oil vapor, mixed with air and exploded in a cylinder to move a piston, provides the power for today's cars and trucks and ships.

As Watt perfected his steam engine a new era was opening, one in which artisans and scientists — or "natural philosophers," as they were called then — began working closely together. Watt became interested in steampower when he was asked by a professor at Glasgow University to repair a broken engine. Collaborations between craftspeople and natural philosophers more inclined to theory would increase rapidly. One of the earliest of these interactions involved the work of a French engineer, Sadi Carnot, and from it emerged the important new field of science called thermodynamics. Though at first concerned entirely with the efficiency of engines, thermodynamics produced the laws that eventually provided answers to such diverse questions as whether particular chemical reactions can occur, and how to attain temperatures close to absolute zero.

Galileo valued what he had learned from talking with artisans at the Arsenal in Venice, a huge armory employing two thousand people — and probably the largest industry of the Renaissance. The key to Venice's fleets and therefore to her domination of world trade, the Arsenal could turn out a standardized galley in one hundred days. According to Galileo,

> conference with [the workers in the Arsenal] has often helped me in the investigation of certain effects including not only those which are striking, but also others which are recondite and almost incredible. At times also I have been put to confusion and driven to despair of ever explaining something for which I could not account, but which my senses told me to be true.[37]

Medieval visionaries such as Roger Bacon had long dreamed of what could be achieved were humans to fathom the laws of nature. Emerging at the close of the eighteenth century, the scientific method, consisting of controlled experiments followed by observation and analysis of the results, was to lead to countless technical achievements. Among them were metals that made possible dreams as fabulous as those of Bacon.

10

The Birth of Modern Metals

> Some [inventions] came to be by the ingenuity of
> those men of speculation, whose trade it is not to do
> anything, but to observe everything; and who, upon
> that account, are often capable of combining to-
> gether the powers of the most distant and dissimilar
> objects.
>
> — Adam Smith[38]

THE LAST THREE CHAPTERS have taken us from the
Roman Era to the Industrial Revolution, from Europe
to China, in order to demonstrate why certain events
happened where and when they did. Paper, printing,
gunpowder, and cannons were invented in China. The
Chinese also produced the first compass, based on the
magnetic properties of iron, and which was so important in their seaborne explorations of Africa and possibly even the Americas, well before the better-publicized
voyages of European mariners. Knowledge of these in-

novations and ideas flowed to the West through Arabia as a result of both trade and conquest. This current went largely in one direction. Chinese emperors were xenophobic, believing that they could learn nothing from the outside world. Once the East's inventions reached Europe, their importance was quickly recognized by Westerners, who began to refine them. These improved technologies were not accepted in China until quite recently, and even then with considerable reluctance.

During the period between the early part of the Common Era and the birth of the Industrial Revolution, the Roman Empire ceased to be the center of power and technological innovation; that center moved to the Near and Far East, where it remained for nearly a thousand years, before returning west again, through Italy, and finally settling in western Europe. Islamic cultures and southern Europe were acquainted with all the inventions from the East. Benefiting enormously from its geographically central location, Italy's merchants had access to western Europe, Byzantium, Arabia, and the Far East. Contact with Syrian glass factories gave Venice, in particular, knowledge of the latest advances in glassblowing. Banking houses developed in Venice and Naples. Gradually, Italy, birthplace of the Renaissance and such technical marvels as the Arsenal, saw its preeminence fade in the 1600s as innovations came out of countries farther north with greater frequency. Italy's commercial importance also waned because gold and silver from Spanish colonies in America were entering through ports along the Atlantic Ocean.

By the end of the Renaissance, a new idea had taken

root and was growing — that it was profitable to turn
out products for the common folk, whose quality of life
consequently improved. A mass market was opening,
one that had the potential of granting great wealth to
those enterprising merchants who could meet its demands. Such entrepreneurs flourished in countries in-
clining toward democratic forms of government, where
they were protected from avaricious monarchs and their
costly follies. A good example of the effect of absolutist
rule on an economy can be found in France during the
latter half of the seventeenth century, when, despite the
enlightened policies of his minister Colbert, Louis XIV's
unproductive and massively expensive wars depressed
trade and helped discourage the spread of manufacturing. By contrast, Great Britain's government was relatively austere, making the nation fertile ground for
industrialization.

Coal for energy and iron for the fabrication of
many important consumer articles were the foundations
of the machine-driven economy characteristic of the Industrial Revolution. Britain possessed both substances
in abundance. But the production of coal and iron in
large quantities required vast amounts of mechanization. Mines needed power for draining water, and for
hoisting and crushing minerals. Iron smelters needed
power for blowing air into increasingly taller furnaces
and for forging increasingly larger blooms. First water
and then steam power satisfied these needs. Large-scale
mechanization became possible only when substantial
amounts of capital were available — provided by banks
and groups of people willing to take risks, if there were

a prospect of large profits. Capitalism was at the heart of all these enterprises.

There had to have also been possibilities for large-scale finance in southern Europe. After all, the earliest banking houses were Italian. In fact, Jacob Fugger visited the House of Tedeschi in Venice in the fifteenth century to learn the business of commerce, triggering the rapid growth of his family's firm in Germany. In both the Low Countries and England the leading banking companies were Italian and Spanish. But the monarchy in England was less absolute than in most other countries in Europe. As we've seen, English sovereigns had no rights to iron and coal mines, and were also constrained from confiscating these properties by institutions such as Parliament. Investors could take risks with the confidence that they would not lose everything to a greedy and arbitrary monarch.

The innovations leading up to the Industrial Revolution were largely the products of Protestant countries. Catholic countries, typified by Spain, had long histories of repressing inquiry. We need only look to the treatment of Galileo by the Church for evidence of its attitude toward the objective pursuit of knowledge, which is a necessary condition for progress in science and technology. Many consider Galileo the father of modern science because of his use of experiment and mathematical analysis of observations. After his early telescopic observations had verified the hypothesis of the Polish astronomer Copernicus — that the earth went around the sun and not vice versa — Galileo came under attack. His conclusion did not fit the Church's earth-centered view of the universe and he was forced to recant his discovery in 1633. Scientists always have a difficult time when they are glancing worriedly over their shoulders.

Under its Most Catholic Majesties Ferdinand and Isabella, Spain expelled its Jewish population in 1492, depriving itself of an energetic and creative middle class. When gold and silver from the New World stopped flowing into its treasury and propping up its economy, Spain's influence on the world scene waned. Intellectuals who perhaps could have found ways to slow or even reverse Spain's decline were thoroughly terrorized by the Inquisition and forced to flee.

Up until the Renaissance, the Church had played a very different role with respect to learning. Instead of being the suppresser of knowledge, it had been the repository of the learning of antiquity. Beginning in the eleventh century, schools and monasteries, through their translations of the scholarship of the Greeks and the laws of the Romans, frequently from Arabic, played a crucial part in the rebirth of Europe.

At the same time that the Catholic Church was having difficulty accepting new ideas, the Protestant Reformation was stimulating inquiry. The essence of the Protestant faith was that any person could approach God directly without the intercession of a priest. Common people were encouraged to read the Bible, which, because of the printing press, was now available to them in German, thanks to Luther's translation. Literacy increased. Thanks to printing, technical innovations could now be quickly transmitted to receptive scholars and artisans. In fact, the Royal Society was founded in England in 1662 for just this purpose. A similar organization, the Académie des Sciences, was established in France in 1665 to foster interaction between science and technology, with one of its first undertakings being a comprehensive compilation of the various crafts in France.

Until the end of the eighteenth century, progress in materials, and in general all areas of technology, came entirely through ingenious craftspeople whose goal was to fashion something of practical value — a tool, a weapon, a new source of power. New materials were discovered in a slow and tortuous, serendipitous fashion. Artisans stumbled (sometimes literally) onto metals, perhaps in a mountain stream or at the bottom of a kiln. People never knew many metals existed until several hundred years ago. Aluminum, nickel, and magnesium, for example, all react so strongly with oxygen in the atmosphere that prospectors would have had no chance to find them in their native state, thus prohibiting their exploitation in antiquity.

Smelters discovered how to extract the less-reactive metals from their ores well before the time of Christ. The key to their success with copper, lead, gold, silver, and iron was the fortuitous happenstance that partially burnt wood is rich in carbon. While charcoal readily reduces iron and copper oxide ores back to their metals, it cannot reduce aluminum oxide ores, because the aluminum-oxygen bond is much stronger than the carbon-oxygen bond. Reduction can be thought of as a sort of competition among elements for bonding with oxygen. The most reactive element wins. Extraction of metallic aluminum and nickel awaited more powerful means of reduction than reaction with carbon, and these emerged with the discovery of electricity.

In its earliest stages, the Industrial Revolution was based on the work of artisans rather than of scientists. Copernicus, Galileo, Kepler, and Newton were natural philosophers, driven by their desire to understand grand and fundamental phenomena, such as how the earth and the other planets circle the sun, rather than by any interest in solving more mundane and practical problems, such as how coal could be used to smelt iron. By the middle stages of the Industrial Revolution, however, the influence of natural philosopher–scientists on technology had increased dramatically, particularly through their efforts to improve the efficiency of the steam engine. Most new materials we will now consider were discovered or at least first identified by those scientists who began calling themselves "chemists."

Up to this point, we have discussed relatively few materials. As we approach our own day, however, their number increases dramatically, making the story of materials more complex, and forcing me to make some drastic selections to keep this book manageable. Introducing the remaining materials means returning to the beginnings of the scientific method.

In the fifth century B.C.E., the Greek philosopher Empedocles postulated that all matter was made up of four elements — earth, air, water, and fire — and that all substances were composed of a combination of them. Aristotle modified his idea, suggesting that one element could transform into another, which certainly fit the

Stephen L. Sass

everyday observation that liquid water evaporates into a vapor.

Motivated by the successes of smelters who transformed minerals into metals, alchemists before the time of Christ pursued their quixotic quest, striving to achieve their holy grail: turning dross into gold. Their efforts came to naught. (Though we shouldn't disparage their follies. After all, malachite, a green stone, yielded copper, a yellowish-red metal.) By the seventeenth century, however, alchemists had come to accept the harsh lesson of nature: to extract gold from a mineral it had to be there in the first place.

In the eighteenth century a fundamental theory of combustion involved "phlogiston," a fiery substance supposedly possessed by materials. When wood or metals burned, went the theory, they gave off phlogiston, leaving behind ash called calx. Curious about the composition of "good" air, Joseph Priestley, an English scientist and Unitarian minister, was the first person to isolate what we know today as oxygen. Focusing sunlight through a magnifying glass, Priestley heated calx of mercury in a closed container and noticed that liquid mercury formed. The gas it gave off supported the burning of a candle. Priestley had discovered oxygen, though he did not understand what happened in this reaction.

When word of what Priestley had done reached France, it piqued the curiosity of Antoine-Laurent Lavoisier, who was also experimenting with gases. Lavoisier spent his days working as a royal tax collector to support his family, but labored in the evenings and on weekends in his private laboratory, his true love.

183

Quantitative measuring devices such as balances were emerging from the shops of instrument makers at that time, and scientists like Joseph Black, a professor at the University of Glasgow and a friend of James Watt, and Lavoisier applied them to weigh the reactants and products of chemical reactions. Lavoisier observed that when a metal was heated in air its weight did not decrease, as should have been the case if phlogiston was given off during combustion. Instead, its weight increased. This observation did not immediately disprove the prevailing theory of combustion, because chemists explained it away by assigning phlogiston a *negative* weight — an example of the difficulty scientists sometimes have giving up on a pet theory.

Meticulously following Priestley's example, Lavoisier heated mercury in air to produce calx and noticed that a candle could not burn in the gas that remained. He then repeated Priestley's experiment, decomposing calx back to mercury and producing a gas that caused a candle to burn even more brightly than in air. Meanwhile, in England, Henry Cavendish had just discovered hydrogen, "flammable air," which he exploded in air to form water. Lavoisier also repeated Cavendish's experiment and then decomposed the water by letting it run down a red-hot iron gun barrel. In a classic series of experiments, Lavoisier showed conclusively that a new gas must exist for combustion to occur. Lavoisier called this gas "oxigene," from the Greek for "acid maker," because he thought that it was present in all acids. Lavoisier was right about the role of oxygen in burning, but he was wrong about it being a constituent of all acids. There are many that do not contain oxygen — hydrofluoric acid, for example. Combustion was now seen as a reaction with oxygen. Lavoisier also saw human breathing as a slow form of combustion. His experiments, which put the phlogiston theory to rest, laid the foundation for the new science of chemistry.

Because he worked for the king, and despite his renown, Lavoisier perished beneath the blade of the guillotine during the French Revolution. The Reign of Terror ended a few months too late for him. A French mathematician, Joseph-Louis Lagrange, remarked that a second's work by a blade had taken the life of a scientist whose like might not be seen again for a century.

At the end of the eighteenth and beginning of the nineteenth centuries, chemistry, like all new sciences, went through a period of organization and classification. Lavoisier and his colleagues in France, and John Dalton in England, searched for compact notations to describe newly discovered gases and fundamental substances such as copper and iron, together with their chemical compounds and reactions. Dalton believed that all matter was made up of very small indestructible particles, or atoms, an idea that can be traced back to Greek philosophers before the Common Era. From such ideas came the fundamental principle that in a chemical reaction the weight of the reactants equals the weight of the products. In other words, matter can be neither created nor destroyed.[39] A Swedish chemist, J. J. Berzelius, devised the classification scheme we use today for the individual elements, in which each is represented by a symbol based on its Latin or latinized name, with Cu (from cuprum) for copper, O for oxygen, C for carbon, and so forth.

One way to classify the various metallic elements is by how readily they react with oxygen. Thus, metals such as platinum and copper are low on the reactivity scale, since platinum has no stable oxide (which is why it is considered "noble" and is frequently found in its native state), and copper has a relatively unstable oxide (which is why it is easy to smelt), while metals such as aluminum and magnesium are high on the reactivity scale and form stable oxides. The element carbon lies between these two groups. A more reactive free element will reduce a less reactive metal in a compound back to its metallic state. Carbon readily reduces copper oxide back to copper, but it cannot reduce aluminum oxide back to aluminum, except under extraordinary circumstances. The ability of one element to reduce another is illustrated by an experiment performed in most high school chemistry classes, when a copper penny is put in a solution of silver nitrate in water. Since copper is more reactive than silver, it replaces the silver in the solution, so that metallic silver deposits on the dissolving copper penny.

At the high-reactivity end of the scale, sodium and potassium can reduce aluminum compounds. This, in fact, is how metallic aluminum was first extracted from its ores during the nineteenth century. How were metallic sodium and potassium first obtained? To answer this, we have to examine electricity, which scientists first became aware of in the eighteenth century. As background to understanding the nature of electricity, which involves the movement of charged particles, we need to consider the constitution of an atom.

At its center, an atom has a nucleus containing protons, positively charged particles, and neutrons, uncharged particles. A cloud of negatively charged electrons surrounds the nucleus, to balance its positive charge. The farther an electron is from the nucleus the weaker its bond to the atom, since the strength of the interaction between charged particles varies inversely with their separation. So if d represents the separation between charges, then the expression $1/d$ describes the interaction. If an element is highly reactive, then it tends to lose its outermost electrons rather easily.

When a compound such as $AgNO_3$, silver nitrate, dissolves in water, it does so by breaking apart. The Ag atoms transfer electrons to the NO_3 molecules, which become negatively charged ions, represented by the symbol NO_3^-. Since they have lost an electron, the Ag atoms become positively charged ions, represented as Ag^+. A water solution of silver nitrate contains Ag^+ and NO_3^- ions moving freely among the H_2O molecules. In the example given above, involving a copper penny in a silver nitrate solution, the copper atom transfers its outer electron to the Ag^+ ion, converting it back to a neutral atom represented by $Ag°$, the silver deposit on the penny. The English scientist Humphry Davy was the first to point out that chemical reactivity and electrical activity are two aspects of the same phenomenon.

With this as background the relationship between electricity and aluminum will become clear. Not found free in nature, since it is far too reactive, aluminum as a consequence is also difficult to reduce from its compounds. The inexpensive extraction of aluminum was, curiously enough, directly linked to the invention of the electric battery by Italian physicist Alessandro Volta.

In 1800, Volta, taking advantage of the different electrical activities of dissimilar metals, devised the "voltaic pile," or electric battery. He put two different metals (each called an electrode) into a conducting solution, connected them together by a metal wire, and noticed that something (what we call a current of electrons) flowed through the wire from the more reactive to the less reactive metal. If the electrodes were copper and silver, the driving force for the chemical reaction described above — involving a more reactive copper atom, which binds its outer electrons more weakly than a less reactive silver atom — causes electrons to move along the wire from the copper to the silver electrode. All batteries operate on this principle. When a charged particle moves, an electrical current results. The difference in chemical reactivity between copper and silver can be thought of as generating a force, measured in volts, which is related to the difference in how strongly the outer electrons are bound to the two metallic atoms. An everyday example is a flashlight battery, which generates 1.5 volts. However, just as in the case of the steam engine, a force must move a distance to do mechanical work. Voltage by itself does no work. The two electrodes must be connected by a conducting wire so electrical current can flow and do electrical work.

As soon as Humphry Davy learned of Volta's invention he set out to use a voltaic pile to decompose various chemical compounds back to their component elements. First Davy melted sodium hydroxide and then passed a current through it, observing that oxygen, hydrogen, and a new soft metal, sodium, appeared. Similarly, potassium hydroxide separated into oxygen, hydrogen, and another soft metal, potassium. Davy had discovered a powerful method of reducing reactive metals in chemical compounds back to their metallic state. His experiments laid the foundation of a new electrochemical industry, which manufactures both metals and gases, such as hydrogen and oxygen. So reactive are sodium and potassium that they can reduce aluminum compounds back to metallic aluminum. This was first accomplished in 1825 by a Danish scientist, H. C. Oersted. Because sodium or potassium are needed to extract aluminum, for many years the reduction of aluminum salts was an expensive process, carried out more on a laboratory basis than on a scale large enough to have a major impact on technology. Sixty years were to pass from the discovery of aluminum to when it could be extracted cheaply and made widely available. While still a curiosity, aluminum was fashioned into knives, forks, and spoons as part of a dinner service for Louis Napoleon in 1848 and was also used to cap the top of the Washington Monument in 1884.

Aluminum has remarkable and desirable properties. A silvery metal, aluminum melts at 660 degrees, well below the temperatures required for copper and iron, making it easier to cast. Its density is low, 2.7 grams per cubic centimeter, as compared to 8.96 for copper, 7.88 for iron, and 19.32 for gold. In the mid-twentieth century, its low density, combined with new strengthening mechanisms, made aluminum the material of choice for aircraft fuselages and rocket bodies where weight was crucial. Aluminum has high thermal and electrical conductivity, and reflects heat and light well. Surprisingly, it is also resistant to corrosion and oxidation. The surface oxide that forms rapidly in the presence of oxygen is continuous, adherent, and very stable, so that once a thin oxide layer forms, additional reaction with the underlying aluminum metal stops.

Despite all these attractive properties, aluminum could never be widely used until an inexpensive process was devised to smelt its primary ore, bauxite, a reddish mineral composed of aluminum oxide and water. Aluminum was transformed from a rarity to a commonplace metal when electricity was applied directly to these ores instead of first being used to obtain potassium, which was then used to reduce the bauxite. An obvious approach would be to melt the oxide ore and then pass an electrical current through it, as Davy did for sodium and potassium. The problem with this method is that because the aluminum-oxygen bond is so strong the melting point of the ore is above 2000 degrees, presenting numerous practical difficulties. How could the high temperature needed to melt the ores be attained and then how could the molten liquid be contained? And what kind of electrodes could survive these harsh conditions? The problems seemed hopeless. But in 1886, two scientists, Charles Hall in the United States and Paul Heroult in France, arrived simultaneously but independently at an ingenious way of extracting aluminum: the electrolysis of aluminum oxide, or alumina, dissolved in cryolite, a mineral containing aluminum, sodium, and fluorine, found in large quantities only in western Greenland. Hall and Heroult dissolved alumina in a low-melting aluminum salt, making it possible to operate an electrolysis cell at the reasonable temperature of 950 degrees. In 1856 aluminum cost $90 per pound, but by 1886, thanks to the Hall-Heroult process (as it is known today), its price plummeted to thirty cents per pound. Just over the horizon were aluminum foil, aluminum beer cans, and aluminum aircraft.

Before the Common Era, large decreases in the cost of materials occurred over many hundreds of years (iron is a good example); only sixty years passed from the discovery of aluminum until the point at which it could be produced economically. Today, the interval between discovery and large-scale production is frequently measured in years rather than decades or centuries, as we will see when we examine polymers. This dramatic contraction in the amount of time needed to develop and market a new material or device is characteristic of our world. World production of aluminum jumped from 2.5 tons in 1873 to 7,300 tons in 1900. It is now the second most widely used metal, with close to fifteen million tons produced annually.

But we need to take a step back. Electrolysis could be used to manufacture aluminum in the quantities I've mentioned *only* if an inexpensive source of electricity were available. The first electrical decomposition involved batteries, which were not economical for commercial-scale extraction of aluminum because their electrodes required large quantities of other metals. The nagging question was, how to get electricity cheaply?

Curious amateurs like Benjamin Franklin — with his kite and key — had experimented with electricity in the eighteenth century. But it was in the nineteenth century that scientists made breakthroughs in the understanding and use of this new phenomenon. In a landmark experiment in 1831 at the Royal Institute in London, Michael Faraday demonstrated that moving a metal wire through a magnetic field induced an electrical current in the wire. Faraday had discovered the basis of the electrical generator, in which the mechanical energy from a steam engine or a water wheel could be converted into electrical energy simply by moving a metal wire in a magnetic field. Thanks to this simple idea, ample electrical power could now be generated to run the Hall-Heroult electrolysis cell, circumventing the need for costly batteries. Faraday also demonstrated that passing a current through a wire immersed in a magnetic field caused the wire to move, the basis of the electric motor.

These seminal experiments sparked a technological revolution. Electricity could be generated at one location, transported via metallic wires, and then used to run an electric motor and do work elsewhere. The energy generated by falling water could now be shipped from the waterfall to where it was needed to do work. At the time of Faraday's death in 1867 the major uses of electricity, such as passing it through a thin wire in an evacuated glass bulb, causing it to glow — the incandescent lightbulb, along with the reduction of aluminum ores and the powering of electrical motors, lay in the future. Perhaps the most important use of electricity during Faraday's lifetime was for the telegraph, making possible for the first time instantaneous long-distance communication. Batteries were the source of power. A young telegraph operator named Thomas A. Edison used discarded batteries, which he stripped of their platinum components, in his search for a filament for his lightbulb. Edison finally hit upon using carbonized thread for the filament in 1879, and then went on to invent or improve upon the phonograph, the electric generator, the carbon microphone for the telephone, and the motion picture camera.

But to return to aluminum. It has a face-centered cubic atomic structure, like copper. When pure it is very weak — again like copper — meaning its yield stress is quite low. Today, of course, one of its most important applications is in airplane manufacturing. How was aluminum strengthened so that aircraft designers could take advantage of its low density? Solid-solution strengthening and work-hardening, processes that worked for copper, also worked for aluminum. As we've seen, strengthening metals means putting up barriers to the motion of dislocations. But in addition to those that we have already discussed, other kinds of barriers exist. These involve hard microscopic particles. For dislocations to move they must either cut the particles or go around them. At issue is how to introduce them into aluminum.

When one substance dissolves in another, the way sugar does in water, more goes into solution at high temperatures rather than at low. A solution containing the maximum amount that it can dissolve at a particular temperature is saturated. When a sugar solution, saturated at 100 degrees, is cooled to room temperature, it now contains more than can be kept dissolved and becomes "supersaturated." The excess precipitates as crystals. This, in fact, is how raw brown sugar is purified: by dissolution at high temperature and crystallization at low temperature.

Exactly the same process occurs within metallic alloys. Aluminum can dissolve copper into a solid solution, where the copper atoms randomly replace aluminum atoms in its face-centered cubic structure. For example, four grams of copper can dissolve in 96 grams of aluminum at 500 degrees, producing an aluminum alloy. If this alloy is then cooled to a lower temperature, for example to 150 degrees, the solid solution becomes supersaturated and the excess copper precipitates in the form of small particles. To better control the microstructure, the supersaturated solid solution is typically first cooled rapidly, or quenched, to room temperature and then reheated to the desired "aging temperature," at which the thin platelike copper particles can be uniformly distributed on a fine scale and remain suspended inside the solid. This is similar to what was done to form martensite in steel, but the end product is quite different. With steel, martensite forms during the quench, while with an aluminum alloy containing four weight-percent copper, a supersaturated solid solution is the result. No change in microstructure occurs until the alloy is aged.

Stephen L. Sass

These tiny copper-rich particles are excellent barriers to the motion of dislocations, which must either cut the particles or go around them. The yield stress of an aluminum-copper alloy containing such precipitates can be 50 to 100 times larger than the 1,000 pounds per square inch for pure aluminum. This strengthening technique is called "precipitation," or "age hardening." Together with the formation of martensite in steel, these are the most important processes for fabricating high-strength metal components.

Aircraft designers rely on aluminum alloys for the fuselages of modern jet planes, because a critical concern is the ratio of the strength of the metal to its density — its specific strength — since the less the body of the aircraft weighs the more payload, that is, people and cargo it can carry. Though aluminum has approximately half the strength of steel, since its density is one-third that of steel, its specific strength is still at least 50 percent greater. Commercially pure aluminum today sells for less than one dollar per pound, while aircraft alloys containing strengthening elements such as copper are several dollars per pound. Aluminum is difficult to surpass when both specific strength and cost are important.

As aluminum came into widespread use in aircraft, designers took advantage of these new high-strength alloys to make fuselages lighter. Unfortunately, a new source of catastrophic failure appeared in components even when used below their yield stress. The failure is due to a special kind of loading, the kind that occurs when the fuselage is pressurized and depressurized during takeoff and landing. This failure process — "metal fatigue" — is a major source of concern for aeronautical designers, not to mention passengers.

But why should a metal component loaded *below* its yield stress fail? The answer is that just as with glass, defects in the form of cracks are always present in metal structures, whether large or small, including bridges, ships, and airplanes. Despite the great care most manufacturers take, scratches or cracks still get introduced either externally on the surface of parts during machining or handling, or internally during casting. And even if by some superhuman effort no cracks were introduced, nicks inevitably occur in components exposed to the environment during everyday use. If the applied load is less than the yield stress, a crack one-thousandth of a millimeter in length will not cause immediate failure, because it is too small to change the stress inside the component. Cracks cause catastrophic failure only when they grow to a critical dimension, perhaps one millimeter in length. A crack originally one-thousandth of a millimeter in length (that's 1/50th the diameter of a human hair) can grow to a length of one millimeter when an aircraft is repeatedly pressurized and depressurized. Cracks in metal components tend to increase in length by perhaps two thousandths of a millimeter per cycle. A back-of-the-envelope calculation shows that under these circumstances a crack achieves the critical length of one millimeter after 5,000 cycles. At that point it is poised to propagate rapidly across the component, a time bomb waiting to explode.[40]

Disasters caused by metal fatigue include the two

crashes of the first commercial jet airliner, the de Havilland Comet, in 1954. Conceived by British designers after the Second World War, the Comet flew at nearly three times the speed of propeller-driven passenger planes and represented the last word in aircraft design, capable of carrying passengers in comfort at altitudes over 30,000 feet. In January and then April of 1954 two Comets crashed following takeoff from Ciampino airport in Rome. The first crash site, near the island of Elba, was accessible to deep-sea diving and recovery, so British engineers were able to reconstruct the shattered fuselage of the jetliner. They discovered that fatigue cracks had formed around a rivet hole in the vicinity of the squarish corners of a window. Sharp corners acted as stress concentrators. Airliner windows today have oval shapes.

At the time of these crashes, the commercial airline business was in its infancy and the pioneering British aviation industry had a well-earned three-year lead over its American counterpart. Alas, public confidence in the Comet plummeted after these accidents, giving American manufacturers like Boeing the opportunity to catch up to and pass the British. Boeing now dominates the commercial aircraft business, particularly on long transoceanic routes with their 747, though over the past decade the Airbus series of jet liners, built by a European consortium, has become competitive. In 1989 the American aerospace business had net exports of $29 billion, twice that of any other industry. The Comet accidents were both tragic and a national financial disaster for the British.

*   *   *

Aluminum is the most common metal in the earth's crust. We turn now to one of the rarest metals, platinum, which though available only in tiny traces plays a crucial role in our world. Platinum was discovered rather late in the scheme of things, and in a part of the world heretofore little explored in this book. While silver and gold were mined throughout the ancient world, platinum, outside of a few specks and strips recovered by archaeologists in Egypt, was initially found only in an isolated river valley in South America — the Choco region of present-day Colombia. Artisans in Choco were the first to work platinum into rings, bracelets, and earrings. Eleven centuries ago, Indians in neighboring Ecuador crafted a beautiful mask of platinum on gold, now in Berlin's Folk Museum. Platinum first came to the attention of Europeans when Spanish conquistadors in pursuit of gold settled in Choco at the end of the seventeenth century. While panning in the local rivers the Spaniards discovered to their dismay that the gold was intermixed with a heavy gray-white metal. This substance was useless to them because it was brittle and their metalsmiths couldn't shape it by hammering, though local artisans had done so for hundreds of years. Frustrated miners called these crystals *platina* — "little silver" — and they abandoned several mines because of the difficulty of separating gold from platinum. Ancient legends tell of people throwing platinum back into rivers in the hope that with time it might mature into gold.

Platinum is silvery white and a "noble" metal because, like gold, it does not corrode or tarnish. Its high density — 21.45 grams per cubic centimeter, which is higher even than that of gold — led to suggestions that it be used as weights for balances and clocks. Like gold, it dissolves in aqua regia ("royal water"), a mixture of nitric and hydrochloric acids. Chemists noticed that once the platinum had dissolved, it always left a residue ranging in color from red to orange to yellow. Encouraged by the French Académie des Sciences to experiment with platinum, Lavoisier tried melting it using an enormous magnifying glass with a four-foot diameter lens and a ten-foot focal length.[41] Though this lens had succeeded in melting iron at 1535 degrees, it failed with platinum, which becomes molten at 1772 degrees. By directing a jet of oxygen gas, which he had recently isolated, onto a hollowed-out chunk of burning charcoal, Lavoisier finally succeeded in liquefying platinum. Later, another French scientist, Guyon de Morveau, purified platinum without using oxygen. First he alloyed it with arsenic and potash to take advantage of the lower melting point, characteristic of an alloy as compared to its pure constituents. Then he fired it to oxidize (and get rid of) the arsenic, yielding a glistening globule of platinum. Finally, de Morveau added salt and charcoal to the platinum, ridding it of all impurities and leaving a ductile metal easily shaped by hammering.

The successes of French chemists toward the end of the eighteenth century fired the imagination of fine jewelers. Best known of the craftspeople who worked this new and exotic metal was Marc-Etienne Janety, goldsmith to Louis XVI, who crafted a sugar bowl from platinum in 1786. Because platinum was found only in

Spanish colonies in the New World, the Spanish king, Charles, sought to restrict its export. Charles's attempt to maintain a monopoly notwithstanding, several hundred pounds of platinum were smuggled into England, where William Wollaston and Smithson Tenant made it the subject of intensive study. During the first few years of the nineteenth century, Wollaston analyzed the residue left over after platinum was dissolved in aqua regia, and demonstrated that the rainbow of colors came from several new metals: palladium, rhodium, iridium, and osmium. The latter two have the highest density of any solid on earth — a whopping 22.6 grams per cubic centimeter. One metal in the residue remained unidentified until forty-two years later, when the Estonian chemist Karl Klaus isolated ruthenium. Palladium, rhodium, iridium, osmium, ruthenium, and platinum are collectively called the "platinum group" metals.

While gold is pursued for its beauty rather than its utility, platinum is sought far more for its value to industry than its aesthetic appeal. Should gold vanish from the face of the earth, outside of dismay in financial and jewelry circles, disruption to our life would be relatively minor. Were platinum to disappear the consequences would be catastrophic, because it is crucial for a wide variety of industrial processes.

Platinum acts as a catalyst, helping to synthesize new and useful molecules from such raw sources as petroleum. The various molecules in oil come together on the surface of the platinum particles and react there much more rapidly than they would otherwise.[42] A true

catalyst is unchanged by reactions that occur on its surface. All American cars possess catalytic converters to
decrease their emissions, and platinum is the key constituent, accelerating the oxidation of undesirable carbon monoxide and nitrous oxide gases left over from the
incomplete burning of fuel in the internal combustion
engine. In the early 1980s, one-quarter of the world's
production of platinum went into catalytic converters.
The effectiveness of a catalyst is easily destroyed (or
"poisoned") by the presence of small amounts of impurity atoms on its surface. Since lead poisons platinum catalysts, unleaded gasoline must be used in cars
equipped with catalytic converters.

Up through the nineteenth century, fertilizers and
explosives depended upon nitrate compounds extracted
from mineral deposits in Chile and guano accretions
(bird droppings) in Peru. Early in this century it was becoming evident that these natural sources were on the
verge of exhaustion. Motivated by their desire to guar-
antee the supply of munitions for their armies, the German chemists Wilhelm Ostwald and Eberhard Brauer
developed a chemical process to synthesize nitric acid,
which can be converted into nitrates. Today, explosives
and nitrate fertilizers are produced almost entirely from
nitric acid, which is manufactured with platinum-
rhodium alloy catalysts. A mixture of ammonia and air
is passed over a platinum-rhodium wire gauze to synthesize the gas nitrogen oxide, which is then dissolved in
water to form nitric acid. Ammonia typically reacts
slowly with air, but in the presence of platinum-rhodium

the reaction rate is a million times faster. As we shall see in chapter 12, platinum also came to play a key role in the manufacture of rayon, the first synthetic textile fiber.

Silver and gold were traditionally valued far more highly than platinum. In fact, during the eighteenth century, counterfeiters gilded platinum coins to simulate Spanish gold pieces. In the nineteenth century, when sterling silver became fashionable, platinum was substituted for silver to fashion less expensive tableware. Today platinum is more costly than both gold and silver: one hundred times more expensive than silver, slightly more than gold. Since Europeans first learned of platinum three centuries ago, it has gone from being a nuisance to gold miners to being absolutely essential for sustaining our world. Its future role is limited only by its scarcity and the imaginations of scientists and engineers.

11

Steel: Master of Them All

Gold is for the mistress — silver for the maid —
Copper for the craftsman, cunning at his trade,
"Good!" said the Baron, sitting in his hall,
"But Iron — Cold Iron — is Master of them all."

— Rudyard Kipling

I RETURN TO IRON AND STEEL because though they played an important role in the technology of Europe and the Near East during earlier centuries, they have literally shaped our own. Steel deserves its own chapter, if only to emphasize its importance compared to other modern metals.

We remember that steel was known to blacksmiths during the first millennium before the Common Era, but they could only fashion thin implements — daggers and swords — because of the difficulty involved in dissolving carbon into the iron. Up to the last third of the nineteenth century, steam engines, railroad locomotives and the tracks they ran on, ships, and bridges — all of them characteristic of the Industrial Revolution — were fabricated from either wrought iron or cast iron, both relatively weak compared to steel. On busy train lines in Britain in the mid-nineteenth century, tracks had to be rotated every three to six months because they changed shape as the locomotives rolled over them. Rails of steel, however, could survive without plastic deformation fifteen to twenty times longer. Boiler walls for steam engines could also be made much thinner (and lighter) with steel than with iron. Weight was important for ships and railroad locomotives, and an urgent need for large quantities of cheap high-strength steel grew rapidly.

Over the centuries, ironmasters seeking to decrease the cost of iron experimented with its smelting and working. The predecessor of the modern blast furnace originated in Catalonia, in Spain, and used hand-operated bellows to provide air. As the smelting ovens in Europe grew taller, waterpower increased the strength of the air blast. Even more powerful blasts were made possible with the advent of the steam engine. John Wilkinson, who was the first to bore cast-iron cylinders for Watt, was also quick to install one of his early engines to service even larger furnaces. In addition, steam was used to power forging hammers and to lift heavy loads. Foundrymen came to realize that the temperature of the furnace could be raised, and smelting made more efficient, by using hot air. The introduction of the hot blast tripled fuel efficiency. By 1850, iron output grew so rapidly that the smelters and forges of Britain, the world's leading producer, were turning out 2.5 million tons annually.

With so much happening in iron production, there were very few comparable innovations in steelmaking. As the middle of the nineteenth century approached, annual steel output stagnated at 60,000 tons. Until the end of the eighteenth century, smiths did not even know that to become steel iron must contain carbon. In 1722 René Antoine de Réamur had described the formation of steel based on the presence of "sulfur and salt." If "carbon" is substituted for "sulfur and salt," he was correct. In 1786, the French scientists Vandermonde, Berthollet, and Monge were the first to claim that carbon was the key to steel. Subsequent early nineteenth-century experiments, in which iron and diamonds were heated together, conclusively demonstrated the need for carbon in steelmaking. As a consequence of all this, the carbon content was erratic, and so were the properties of the steel. Steel was still made by "cementation" — heating thin pieces of iron by placing it on charcoal for periods of up to ten days. During the eighteenth century, the English clockmaker Benjamin Huntsman developed the crucible process for steelmaking. First, high-carbon pig iron, so-called because the arrangement of the molds reminded ironworkers of a litter of suckling pigs around a sow, was decarburized by heating. Then approximately fifty pounds of the resulting wrought iron were reheated in special clay crucibles to reintroduce a sufficient amount of carbon to make steel. This process typically took ten days to transform iron to steel — not very promising for large-scale production.

The first economical approach to the mass production of steel emerged nearly simultaneously in the United States and Britain. In the 1840s, British and Prussian arms manufacturers were plagued by their experimental weapons exploding upon first firing, because of the varying composition and microstructure of crucible steel, coming as it did from small batches. A brilliant English inventor, Henry Bessemer, pondering ways to fire specially shaped projectiles (as distinct from round cannonballs), quickly realized that steel was the only practical material for such armaments. Taking advantage of the strong reaction between carbon and oxygen, Bessemer devised a process to de-carburize pig iron by blowing a blast of hot air through the molten metal. Large quantities of steel now took hours to produce rather than days.

As word of Bessemer's process spread, ironmasters in Britain flocked to him, and within a month in 1856 he had amassed a fortune in royalties. But success was not to be won so easily. While Bessemer's approach was theoretically correct, in practice it yielded brittle steel. Most English ores, as it turned out, contained phosphorus, which was not removed by his process. All of Bessemer's experimentation had been done with phosphorus-free Swedish ore. An additional problem with the Bessemer process was that frequently it removed all the carbon from the iron, while at the same time introducing bubbles of oxygen. As Bessemer later recalled, this failure was "like a bolt from the blue; its effect was absolutely overwhelming. The transition from what appeared to be a crowning success to one of utter failure well nigh paralysed all my energies."[43]

One of the hallmarks of a successful inventor or scientist is dogged perseverance in the face of failure. Bessemer persisted, frantically working day and night, but without success. A solution finally did emerge, however, though not as a result of Bessemer's desperate labors. Instead, it was produced through the efforts of another English foundryman, Robert Mushet, who appreciated that chemistry was the key to the problem. He added a German compound called *Spiegeleisen,* containing iron, carbon, and manganese, to the molten iron. Manganese reacts with the oxygen in the iron to form manganese oxide, which passes into the slag and is removed, while the carbon is sufficient to raise the alloy content back to the level needed for steel. Bessemer's introduction of Mushet's innovation — which, interestingly enough, he never acknowledged — led to the Bessemer process gaining acceptance first in England, then on the Continent and in the United States.

How large-scale production of steel came about is actually a more complex tale than I'm suggesting here. Nearly a decade before Bessemer, an ironworker in Kentucky named William Kelly also realized that hot air could de-carburize molten pig iron. After much frustration and many financial woes, Kelly finally demonstrated that his idea worked. He applied for a patent after learning that Bessemer had obtained one in America. Kelly's patent was given precedence, but he had little success in defending his work. Though Kelly received royalties of several hundred thousand dollars, the large-scale production of steel came to be credited not to him but to Bessemer, since the first steel rails used in the United States were stamped "Bessemer Steel." Mushet fared even worse, reaping only a few hundred pounds for his innovation, though he reckoned his fair share should have been in the hundreds of thousands of pounds.

Steel production grew exponentially in the United States, climbing from 22,000 tons in 1867 to one million tons in 1880. By 1900 it had reached nine million tons, one-third of the world's output, thanks to the vast quantities of iron ore coming from the newly opened Mesabi Range in Minnesota. Production continued to mushroom, tripling again over the next decade. Germany also experienced explosive growth, its steel output leaping from 12,000 tons in 1850 to eight million tons in 1900. What could rightfully be called the Age of Steel had begun. First the United States and then Germany surpassed Britain in steel production during the last two decades of the nineteenth century. Having dominated international trade in iron for more than a century, Britain's nearly 70 percent share of the world market in the 1870s fell to 30 percent by the outbreak of the First World War. In only a few decades, Britain, birthplace of the Industrial Revolution, had slipped from her position of technological dominance.

First iron's and then steel's dominance continued apace. The first iron bridge traversed the Severn River in Britain in the 1770s. Over the next one hundred years, wrought and cast iron were used for increasingly longer spans, as engineers gained confidence in their material and their designs. Steam engines, and the gears needed to transmit their power, were cast from iron. Builders began using iron plates for ship hulls early in the nineteenth century, and when iron ships survived running aground, such as the *Garry Owen* in 1834, the remaining skeptics were convinced. Steam engines had been introduced into wooden ships in the early nineteenth century as well, and the first successful transatlantic run was made by the *Great Western* in 1838. A short time later, the steam engine was successfully wed to the iron hull in the *Great Britain,* followed in 1858 by the monstrous *Great Eastern,* the largest and fastest ship of her time — double the length and triple the weight of any vessel before her. And then, during the American Civil War, the navies of the great sea powers were transformed by the famous clash of ironclads, the Union *Monitor* and the Confederate *Merrimac,* at Hampton Roads, Virginia. The battle rendered all wooden ships obsolete.

Steel's properties, particularly its higher yield strength, are far superior to those of wrought and cast iron. First the Bessemer process and then the open-hearth furnace process, developed a short time later by Siemens in England and Martin in France, made steel cheaper than wrought and cast iron. Those phosphorus-rich ores, so disastrous for Bessemer, were finally smelted successfully in the open-hearth furnace by utilizing a refractory brick liner, which reacted with and removed the undesirable impurity. Difficulties still remained, however, in the heat treatment of steel. Cooling it too slowly made it weak and cooling it too rapidly made it brittle. A subsequent anneal, or tempering, of the martensite formed during the quench was needed to make the steel more ductile. Optimum strength and ductility of steel required careful control of the heat treatment conditions.

The *Banshee,* a blockade runner built by the British during the Civil War, was the first steel ship, with plates half the thickness of iron, cutting her weight in half. Yet in 1875, Nathaniel Barnaby, chief naval architect of the British Admiralty, pondering the building of steel ships, inquired of the industry, "What are our prospects of obtaining a material which we can use without such delicate manipulation and so much fear and trembling?" The process control achieved with the open-hearth furnace gave the Admiralty a satisfactory answer, and in 1877 it immediately commissioned the first steel warship, *HMS Iris.* This was the launch of a period of rapid growth in steel shipbuilding. Twenty-five years later, at the dawn of the new century, few ships were still being constructed of iron. In less than one hundred years, first wooden and then iron ships had become obsolete, a remarkably fast transition for the navies of the world.

Iron and then steel revolutionized building construction during the 1880s. In antiquity, starting with the tower in Jericho and later the ziggurats in Sumer, the pyramids in Egypt, and the Pantheon in Rome, all substantial buildings were erected with massive lower walls to bear the weight of the upper structure. When walls bear the weight of the building, relatively few windows are put in. They weaken the edifice. In 1883 the Home Insurance Company in Chicago commissioned William

Jenney to erect a fireproof, ten-story building with a large number of windows. Jenney decided to transfer the weight of the upper portion of the building away from its walls to an internal skeleton of iron, which then transmitted the load to the foundation. Relatively thin, lightweight materials could therefore be used for the walls. Their main purpose was to shield the building's occupants from the weather. The human body's construction, with an endoskeleton, a framework of bones bearing our weight, and our skin as the protective layer, served as an excellent model for architects to emulate. What would we look like if we did not have an internal skeleton and had to rely on our skin, muscles, and internal organs to bear the weight of our body? Very differently than we do. Walking upright would be challenging. We would likely have evolved with an exoskeleton, an external skeleton, like a lobster's.

Together with Elisha Otis's safety elevator, Jenney's innovative design ushered in the era of skyscrapers. Elevators before Otis's day were plagued by accidents, since the passenger cage hung from a long rope, which eventually became worn and broke. In 1854 Otis demonstrated his safety elevator, featuring ratchets along the walls of the shaft that were designed to catch a falling cage. Jenney's Home Insurance Building was erected to a height of ten stories. Wrought iron was used for the lower portion and Bessemer steel girders for the upper four floors. The greatest monument to wrought iron was the 985-foot Eiffel Tower, symbol of the World's Fair in Paris in 1889. Designed by the French engineer Gustave Eiffel, who also constructed the framework for the

Statue of Liberty, and made with 7,300 tons of iron, it was for many years the world's tallest structure. Rand-McNally in Chicago then commissioned the first entirely steel building in 1890. New Yorkers, recognizing the innovations of their prairie brethren as well as the financial advantages of such buildings in a time when land was becoming scarce, soon transformed the skyline of Manhattan with skyscrapers.

Steel also revolutionized bridge construction. In 1890, a massive span was built across the Firth of Forth in Scotland. Iron cables had already been used as supports for a 702-foot suspension bridge across the River Avon in England. The cables were assembled on land and then strung across the river to the other bank. An American engineer, John Roebling, working on the Brooklyn Bridge, improved on this by assembling the cables in place, drawing the steel wires strand by strand across the East River. By the turn of the century, large cities, particularly in the United States, had taken on strikingly new appearances because of steel.

Giving steel proper credit for its role in shaping our modern world requires a book of its own, and there are many. The domination of iron and steel in all manner of construction is demonstrated by the huge quantities produced today. The world's output in the past few decades has surpassed a half billion tons annually. The ability to shape ingots by cold-rolling contributed to the growth of vast new industries, including automobile manufacturing, in which car bodies were and still are formed out of thin steel sheets. In addition, toward the end of the nineteenth century, steels were no longer formulated simply from iron and carbon. As early as the time of Michael Faraday, foundrymen realized that adding other elements produced alloy steels with vastly improved properties; for example, nickel and chromium greatly enhanced corrosion resistance, giving a new class of "stainless" steels. The addition of tungsten produces extremely high-strength steel, while adding manganese makes it possible to form martensite without having to quench steel in water, which is impossible to do for thick components.

Toward the end of this century the United States fell behind in steel technology. Innovations now come primarily from Europe and Japan. Remarkably enough, the United States Steel Corporation, for so long a name synonymous with the steel industry, dismantled its research laboratory in the early 1970s, changed its name to USX in the 1980s, and then considered selling off the remaining portion of the company devoted to steelmaking in the 1990s. The history of iron and steel illustrates the advantages that accrue to those who relentlessly pursue technological superiority, and how quickly the mantle of innovation passes from one country to another. Production of iron and steel in the United States fell from a peak of around 220 million tons per year in 1970 to 120 million tons by 1986. The United States overtook Great Britain in steel production at the end of the nineteenth century. A century later, world leadership has moved on to other countries such as Japan. Will the United States in the twenty-first century follow the same path Britain did in the twentieth? Perhaps a more important question is whether steel production remains the best indicator of technological dominance. Could newer, more exotic materials now be assuming that role? Examining polymers, which, despite their abundance in nature and in the human body, were first identified as a distinct class of materials only in this century, might provide an answer.

12

Exploding Billiard Balls and Other Polymers

> There is probably no other inert substance the properties of which excite in the human mind an equal amount of curiosity, surprise and admiration. Who can examine and reflect upon this property of gum-elastic without adoring the wisdom of the Creator?
>
> — Charles Goodyear (1855)[44]

"SUPPOSE THAT YOU WERE COMPARING artifacts dredged up from the wrecks of two ocean liners, the British *Titanic,* which went down in 1912, and the Italian *Andrea Doria,* which sank in 1956. What do you think would be the most surprising difference between what you found on those two ships?"

My family and I were driving across Idaho on one of our trips to Twin Falls to visit my in-laws, and we were passing the time pretending we were contestants on a game show. It was my turn to try to stump the experts, and I was confident that now I had done it. First my sons went through a collection of morbid relics, human and other, retrieved from the deep, reflective of the interests of young boys: parts of human bodies, sharks that choked on a shoulder blade, and so on.

"Wrong," I kept repeating.

Suddenly my son Erik glared at me.

"Dad, I know you. This has to do with materials."

Trying not to smirk, I raised an eyebrow and nodded. Both Adam and Erik shouted out the answer simultaneously: "Plastics!"

They were right. We didn't learn to make wholly synthetic materials until this century and few, if any, existed in 1912. By 1956 we were already very familiar with plastic radios, raincoats, seat covers, and wire coatings. In the 1967 movie *The Graduate*, Mr. Robinson's advice to Benjamin about what to do with his future — "Plastics!" — was actually excellent advice. To Benjamin, of course, plastics embodied everything he was trying to run away from.

Chemistry's role in uncovering new materials emerged toward the end of the eighteenth century, grew in importance with the advent of new reduction and synthesis techniques in the nineteenth century, and continues to expand in the twentieth century. Which among the vast number of discoveries made through chemistry was the most important is a question sure to ignite vehement arguments, but high up on anyone's list has to be polymers. "Polymer" comes from the Greek *poly*, "many," combined with *meros*, "parts." A polymer molecule consists of a long chain made up of basic building blocks, *mers*. Its properties depend upon the number and nature of mers in that chain. Because some polymers can exist with either amorphous or crystalline structures, they have intriguing and often unique ways of behaving. For example, their mechanical properties can change rapidly with small changes in temperature. Rubber, one of nature's polymers, goes from being very flexible to very stiff over a small temperature interval. Such behavior was the origin of the *Challenger* space shuttle disaster in 1986. Rubber can be both watertight and gas-tight while remaining quite flexible. It is also a good electrical insulator.

Polymers abound in nature. Our bodies are made up of polymers. Our skin — that remarkably flexible protective layer keeping us in and the rest of the world out — brains, muscles, nerves, and nails are all composed of long chain molecules. DNA, the famous double helix molecule, consists of two chains coiled around each other. Wood, containing natural polymers, was for millennia the primary material for fabricating houses, boats, carts, and parts of tools and weapons. Silk, which is a polymer fiber unwound from the cocoon of the silkworm caterpillar, was first woven into textiles by the Chinese. Legend has it that this occurred as early as 2700 B.C.E., and led to the establishment of the Silk Road.

Within the past century, we have learned how to synthesize polymers, and that knowledge has transformed our lives. Rayon, nylon, Teflon, and Kevlar are but a few of the myriad synthetic polymers to have emerged from the flasks of chemists, affecting most everything we do in our daily life, from the clothes we wear to the movies we watch. Understanding this fascinating substance requires investigating first how humans came to recognize and use the polymers nature provided for us, and then how we learned to emulate, and even improve upon nature.

Rubber is a good place to start. Following Christopher Columbus's voyage to America in 1492 (recognizing that there are other claimants to the honor of discovering it, including the Vikings and the Chinese), he went on to explore the islands of the Caribbean. His journals provide a description of natives in what is today Haiti playing with bouncing balls, one of the first references to rubber. Over two hundred years later, Charles de la Condamine, a scientist in a French expedition to South America, informed the Académie des Sciences in Paris of a novel rubbery material extracted from latex, the milky white juice oozing from cuts in the bark of the *hheve* tree. Following exposure to heat or air over long periods, fine particles suspended in the latex coagulate to form natural rubber, which the Indians in Ecuador and Peru knew as *cahuchu*, or "weeping wood," and the French called *caoutchouc*.

Indigenous peoples fashioned waterproof boots by spreading latex on their feet and letting it dry, and crafted water bottles by coating a clay pot with latex. Because it coagulated so easily, latex could not be shipped long distances, so the Indians traded rubber in the form of large balls. European scientists were quickly drawn to this curious substance, hoping to discover new uses for it. For years, people in Europe used rubber in much the same ways as the South American Indians, until Joseph Priestley, the discoverer of oxygen, observed that it could rub out pencil marks — hence the provenance of the English word for "eraser," "rubber." Even Michael Faraday became intrigued with rubber, and did experiments that proved it contained carbon and hydrogen.

In the 1820s Charles Macintosh, a Scottish chemical manufacturer, specialized in waterproofing textiles by coating them with a solution of rubber in naphtha, lending his name to the British term for raincoat. Rubber clothing and boots had serious drawbacks, however. They tended to become sticky and change shape, or sag on a hot sunny day. Looking for a way to overcome this problem, Macintosh sandwiched a rubber layer between two sheets of fabric to protect the wearer from its stickiness. Rubber also lost most of its elasticity on cold days, becoming hard and brittle, and another difficulty was that solvents such as naphtha dissolved rubber. To understand how these problems were finally solved, some background on the atomic structure of polymers is necessary.

Simple polymer molecules are made up of large numbers of identical mers repeated along a linear chain. A molecule of polyethylene, for example, is represented compactly by the symbol $(C_2H_4)_n$, where n stands for the number of mers in the chain (or the "degree of polymerization"), which can be 1,000 or 10,000 or more. "High polymers" have large numbers of mers.

Carbon atoms have four bonds that when their tips are joined together form a tetrahedron. These types of atomic bonds are strongly directional and are called "covalent" bonds. When one carbon atom is linked to another, their bond consists of two shared electrons. Diamond — which we'll examine in the next chapter — is formed when all four bonds of a carbon atom are linked to other carbon atoms, building up a three-dimensionally periodic structure. In metals, the bonding between atoms is nondirectional. The positive nucleus of the atom is surrounded by a sea of electrons, giving a "metallic" bond. Another kind of nondirectional bond involves atoms with charges of a different sign on their nuclei (ions). They are held together by the electrostatic attraction between negative and positive charges, giving an "ionic" bond. Such bonds exist, for example, between the positive sodium and the negative chloride ions to make sodium chloride, ordinary table salt.

To produce the polymer backbone, carbon atoms link up in a row, connected by two of their four bonds. The remaining two bonds are then available to join up either with individual atoms, such as hydrogen or chlorine, or with larger clusters of atoms (known as "side groups"). Atoms along this chain can rotate freely about the carbon-carbon bond linking them together, making the polymer molecule highly flexible. This is particularly true if small atoms, such as hydrogen, are bonded to the backbone carbon atoms. When larger clusters of atoms are present, they run into one another when the chain attempts to flex, causing it to kink or coil — a process known as "steric hindrance" — steric simply meaning "spatial arrangement." Modern polymer chemists attempt to tailor the structure and mechanical properties of polymer chains by adding appropriate side groups, as we'll see.

Natural rubber's molecular structure is more complex than that of polyethylene. It has a new feature, a double bond between neighboring carbon atoms along the chain, restricting rotation around the carbon-carbon bond. In addition, the methyl side group in rubber is large enough to run into neighboring hydrogen atoms on the chain. Because of this steric hindrance, the polymer chain kinks up. When natural rubber is stretched at room temperature — in other words, put under a load — the individual molecules uncoil. This produces large changes in dimension for small loads, sometimes more than a doubling in length. Eventually, after the individual molecules uncoil and become more or less parallel, they begin sliding by one another, like slithering strands of spaghetti. When the load is removed, the molecules spring back to their coiled shape and the rubber decreases in length. However, the molecules have no memory of their starting point relative to one another before loading, so they can't spring back to their original shape. The rubber remains permanently deformed (or "set"). Natural rubber's elastic behavior is therefore due to molecular uncoiling and coiling, and its plastic behavior stems from the way the molecules slip with respect to one another.

Natural rubber's stickiness and its tendency to sag at high temperatures result from the way its molecules move, then forget where they started. Since temperature has a large influence on the movement of molecules, rubber's mechanical properties change rapidly and dramatically with heating and cooling. At temperatures slightly above room temperature, natural rubber plastically deforms under its own weight, which was why early tires drooped in the hot sun; at temperatures below the freezing point of water it behaves like glass, shattering under hard impacts.

Although the elastic and plastic deformation behaviors described here are superficially similar to those of metals we considered earlier in chapter 2, they are actually quite different, both in mechanism and in magnitude. With metals such as copper, the extent of elastic deformation is far less than 1 percent — large loads produce small changes in length. For natural rubber, however, the change in length can be much greater than 100 percent — small loads produce large increases in length when the long flexible molecules uncoil. In metals, plastic deformation results from dislocations moving; in polymers, when molecules slip by one another. Thus the Young's modulus (or stiffness) of rubber is less than one thousandth that of steel, which is one of the many reasons why metals and not rubber are used for construction. Driving across a rubber bridge that sank twelve inches as you passed over would be disconcerting. On the other hand, driving a car with steel tires would be very uncomfortable. In this case the low stiffness of rubber is advantageous.

In the early part of the nineteenth century, companies selling products fashioned from rubber frequently went bankrupt; customer satisfaction with merchandise that deformed and became sticky so easily was not high.

This was particularly true in the United States, where Macintosh's relatively complex process was not being employed. How could rubber's drawbacks, related as they were to the intrinsic properties of its molecules, be overcome? A hardware dealer in Connecticut named Charles Goodyear was obsessed by rubber and devoted himself to unraveling its puzzles. Beginning in 1832, he experimented with different additions. In 1839, after accidentally burning a mixture of rubber, sulfur, and lead carbonate, Goodyear was picking through the smoldering and charred remnants on his stove when he came upon a chunk of a new substance that had not softened. He had discovered that sulfur joins together (or cross-links) individual rubber polymer chains to produce a three-dimensional network. Double bonds between carbon atoms in rubber molecules are highly reactive and can be broken by sulfur atoms, which then link the two molecules together. Now when a load is applied, the molecules still uncoil, but when they slide by one another the sulfur cross-links between neighboring chains give them a memory of their initial positions. This process came to be called vulcanization (in honor of Vulcan, the Roman god of fire), or curing, of rubber.

Goodyear did not understand why sulfur improved the properties of rubber so dramatically, but he couldn't have cared less. Sulfur magically solved all of rubber's pesky problems. Later experiments by Goodyear's brother, Nelson, showed that as the amount of sulfur was increased the rubber became inflexible, resulting finally in an extremely hard material called ebonite, after "ebony." As the amount of sulfur increased, so too did the number of cross-links between chains, making it even more difficult for the molecules to slide by one another and rendering the three-dimensional polymer network yet more rigid.

Hoping to interest the Macintosh Company of England in his process, Goodyear naively sent them a piece of his cured rubber. It came into the possession of Thomas Hancock, a pioneer in the working of rubber. Intrigued by its properties, Hancock set out to discover their origin. Goodyear delayed applying for a patent until he had perfected his process, since at first only the surface of the rubber was cross-linked, not its interior. Hancock, meanwhile, discovered the role of sulfur in the curing process and applied for a British patent, which he received in 1844 — two months before Goodyear's American patent was granted. Goodyear's British patent was therefore rejected, leaving him to complain bitterly about the injustice of patent law in England. Nevertheless, rubber was now on its way to widespread use. Between 1840 and 1860 rubber production grew from 150 tons to 6,000 tons per year. With the birth of the automobile industry at the end of the nineteenth century, an enormous demand arose for inflated (or pneumatic) tires, which took advantage of rubber's unique combination of flexibility and airtightness to give a more comfortable, less bone-jolting ride than earlier, metal-rimmed wheels. Rapidly increasing consumption led to shortages, which forced the rubber industry away from its dependence on wild trees located in Central and

South America, particularly along the Amazon River, and toward the development of extensive plantations in Malaya, Ceylon, and Indochina.

As the twentieth century opened, the mobility of armies depended entirely on pneumatic tires. Countries vulnerable to being cut off from their supplies in the Far East during time of war searched for ways to produce an artificial rubber. Throughout the nineteenth century, chemists were synthesizing and identifying large numbers of novel carbon-based, or organic, substances. They discovered that heating rubber released a new chemical they called isoprene. Just before the outbreak of the First World War, chemists at the German firm I. G. Farbenfabriken developed the first synthetic rubber, based on isoprene. In its wisdom, however, a German government commission decided that there was no need to build synthetic rubber factories, concluding that the war would be over before they were completed. As expected, once the war started Germany lost access to its foreign supplies because of the British blockade. Its stockpiles dwindling, Germany resorted to smuggling rubber disguised as coffee through neutral countries. Germany eventually manufactured several thousand tons of synthetic rubber.

After the First World War the price of natural rubber fluctuated wildly, from a few cents to over one dollar per pound. Efforts to improve the synthesis of rubber floundered until Hitler, anticipating wartime shortages, prudently pushed the development of a more expensive synthetic rubber called Buna S. During the Second

World War, following the Japanese capture of the vast Far Eastern plantations that had been supplying 90 percent of its natural rubber, the United States embarked on a crash program to increase synthetic rubber production, which went from 9,000 tons in 1941 to over one million tons by 1945. Today, more than eight million tons of synthetic rubber are consumed annually in the world, twice the amount of natural rubber.

We have seen that sometimes one substance can be substituted for another. Brick or wood does as well as stone in house construction. As Germany discovered to its chagrin during the First World War, rubber was different: nothing could replace rubber for use in automobile tires and flexible tubing. Today these two uses alone consume more than 60 percent of America's output. And high-sulfur hard rubber is favored where mechanical strength and corrosion resistance is needed, such as in automobile battery casings.

When it is vulcanized with a small amount of sulfur, rubber becomes what is called an "elastomer," a material that can undergo a large amount of elastic (reversible, in other words) deformation. Rubber bands, which can stretch to six or seven times their original length without plastic (irreversible) deformation, are everyday examples of elastomeric behavior. This property is strongly temperature dependent, since cooling an elastomer by only 30 to 40 degrees frequently transforms it into a glasslike material. When this occurs, the substance is said to pass through its "glass transition" and becomes a glassy polymer. This substantial change in mechanical properties with small change in temperature was the origin of the *Challenger* space shuttle disaster. Large rubber gaskets were used to seal the joints between the four sections making up each of the two external shuttle booster rockets. The gaskets were required to change shape quickly in response to the slight misalignments of the individual sections of the rocket, misalignments caused by the high internal pressures generated by the burning propellant. That fatal winter morning in January 1986 the temperature at the launch site was 36 degrees Fahrenheit, 15 degrees cooler than it had been for previous launches. The temperature at the joint was even lower, 28 degrees, and the consequent sluggish response of the gasket allowed hot gases to escape through the joint, destroying the gasket and eventually igniting the main fuel tank containing the liquid hydrogen fuel. Since then the gaskets have been redesigned, and NASA avoids low-temperature launches.

The next polymer we'll examine also comes from trees, not from their sap but their wood. In 1868 the Phelan and Collender Company of Albany, New York, offered a $10,000 reward to anyone who could come up with a substitute for ivory. Phelan and Collender was in the billiard ball business. Elephant herds in Africa had been decimated for their ivory, which was in short supply. The search for a replacement material for billiard balls would eventually yield rich rewards, including the first motion picture film and the first artificial textile fiber.

For many centuries, wood pulp — bundles of cellulose fibers — was the basis of papermaking. In 1846, Christian Schönbein of the University of Basel in

Switzerland dissolved paper in a solution of nitric and sulfuric acid, producing a new substance, cellulose nitrate. After working on the vulcanization of rubber, a successful British metallurgist named Alexander Parkes experimented further with cellulose nitrate, hoping to discover a way to soften and shape it into an insulator that could be used in the rapidly emerging electrical industry. He found that camphor worked well as a softening agent (or plasticizer) and produced Parkesine. Parkes won an award for his invention at London's Great International Exhibition in 1862, an event carefully designed by Great Britain to flaunt its industrial might. Parkes, a better inventor than he was an entrepreneur, saw his company eventually fail, even though Henry Bessemer was on its board of directors.

Meanwhile, across the Atlantic, the search for billiard ball substitutes also spurred John Hyatt into experimenting with cellulose nitrate. He discovered it could be dissolved in a solution of camphor and ethanol. Cellulose nitrate, or "celluloid," as it was called by John's brother, Isaiah, could now be poured into molds and easily shaped into a wide variety of new products. John and Isaiah manufactured detachable collars and cuffs, children's dolls, combs, dentures, and, eventually, billiard balls. There was one problem. The billiard balls exploded when touched by lit cigars. Cellulose nitrate is synthesized by replacing one of the three hydroxyl ions on the cellulose mer with a NO_3 ion. If all three hydroxyl ions are replaced, cellulose trinitrate, also known as guncotton, results. As with gunpowder, the presence of the NO_3 ions makes large amounts of oxygen available to accelerate combustion. Cellulose nitrate is highly flammable. Guncotton, with three times as many nitrate ions, is positively explosive. Since celluloid could be formed into a thin transparent film with good mechanical properties, it also played an important role in the early days of the motion picture industry. Unfortunately, its flammability occasionally led to disaster, as dramatically portrayed in the film *Cinema Paradiso*, in which the projectionist is blinded by a fire. Celluloid is so unstable that it also reacts with small amounts of impurities in the air, and in time cellulose nitrate films slowly deteriorate. A number of early film classics have been damaged, some to the point of being irretrievable, by this decay.

Cellulose nitrate is a "thermoplastic," meaning a polymer that can be shaped and reshaped by the application of heat and pressure. The generic term "plastic," used to refer to many polymers, actually comes from thermoplastic. Rubber is an example of a polymer that can be transformed from a thermoplastic to a thermosetting material by cross-linking it with sulfur. Thermoplastic polymers have a linear chain structure, while thermosetting polymers have a three-dimensional network structure.

Using cellulose nitrate as a base, chemists turned to synthesizing artificial fibers from cellulose. As early as 1665, the Englishman Robert Hooke speculated in his work *Micrographia* about the possibilities of obtaining artificial silk:

And I have often thought that probably there might be a way . . . to make an artificial glutinous composition resembling, if not full as good, nay better, than the Excrement or whatever substance it be out of which the Silk-worm with-draws its clew. . . . This hint may, I hope, give an Ingenious Inquisitive Person an occasion of making some trials.[45]

In the nineteenth century the ingenious and inquisitive Frenchman Hilaire Chardonnet, building on the work of a series of French and British scientists on the synthesis of artificial silk, realized that extruding cellulose nitrate to form fibers was not desirable because of its high flammability. Chardonnet devised a method of replacing the nitrate ions on the cellulose molecule and began to manufacture artificial silk in 1890. At the beginning of the twentieth century, a new material was produced by extruding cellulose xanthate dissolved in sodium hydroxide into a solution of sulfuric acid and sodium bisulfate. Forcing the thick molasseslike solution called "viscose" through tiny holes in a platinum spinneret — platinum, which as we know is inert, was needed to resist the highly corrosive chemicals involved in extrusion — gave the first artificial fiber, rayon. The era of synthetic fibers had arrived — first came rayon; then in the 1930s, nylon; in the 1950s, Dacron; and in the 1970s, Kevlar. A variety of other products based on cellulose emerged. The thin transparent flexible film named cellophane is perhaps the most ubiquitous.

The early part of the twentieth century saw the synthesis of the first completely artificial material, Bakelite, manufactured not by regenerating a natural polymer, as with rayon, but entirely from chemicals. In a series of experiments that led to over one hundred patents and a small fortune, Leo Baekeland, a Belgian chemist working in the United States, synthesized a new polymer from phenol and formaldehyde. Bakelite, a thermosetting polymer, was formed in a condensation reaction using heat and pressure. Condensation polymerization involves the reaction between two chemicals where hydrogen ions from one join with hydroxyl ions from the other to form water, with the remaining molecules reacting to produce the desired polymer.[46] Bakelite is a good insulator, has reasonable mechanical strength, and has the additional benefit of emerging from the reaction mold in final form. This was ideal for the mass production of electrical components for automotive and household applications. By the end of the Second World War, production of Bakelite and similar phenolic resins reached 125,000 tons annually worldwide. Bakelite's first commercial use was in the gearshift handle of the 1916 Rolls-Royce.

The first chemists of this century were stunningly successful in synthesizing new and intriguing materials, though they had very little idea about the long chain nature of polymers. Indeed, until the experimental work of the German chemist Hermann Staudinger clarified their true nature, there was heated debate over what the molecular structure of polymers actually was. Staudinger received the Nobel Prize in Chemistry in 1953 for his pioneering work on giant or long-chain molecules.

*   *   *

During high-pressure experiments on reactions between ethylene and benzaldehyde at the Imperial Chemical Industries laboratories in England in the 1930s, Eric Fawcett and Reginald Gibson accidentally polymerized ethylene, forming polyethylene, the last polymer I'll discuss, and the simplest of them all. This linear polymer went on to find many uses in containers, tubing and wire, and for cable coatings. According to Robert Watson Watt, the British inventor of radar, polyethylene also played a crucial role during the Second World War. "The availability of polyethylene transformed the design, production, installation, and maintenance of airborne radar from the almost insoluble to the comfortably manageable," he wrote. "And so polyethylene played an indispensable part in the long series of victories in the air, on the sea and the land, which were made possible by radar."

Because of the flexibility of their coiled molecular chains, most linear polymers have relatively low Young's moduli (and stiffness) and therefore are not useful in applications in which elastic deflections must be small. When polymers are crystallized, however, by stacking their straight chains in an orderly, periodic fashion, their stiffness increases enormously, because now deformation cannot occur by the uncoiling of long molecules but instead by the stretching of bonds along the carbon-carbon backbone. High-pressure reactions of the type discovered by Fawcett and Gibson assembled polyethylene molecules with branched chains, which do not pack well together and cannot be easily crystallized. Nature seems to have solved this problem, because polymers such as rubber have unbranched chains.

Stephen L. Sass

In the early 1950s a revolutionary breakthrough in polymer synthesis occurred, one that materials scientists like myself still look back upon with awe. Working at the Max Planck Institute for Coal Research in Mulheim, Germany, Karl Ziegler produced straight polyethylene molecules, with as many as 100,000 mers in a chain, at atmospheric pressure. The secret to his success was a catalyst, first containing nickel, then zirconium, and finally titanium, which both accelerated the polymerization reaction and for some unknown reason favored only straight polymer chains. This was the first highly specific catalyst to assemble polymer chains just one way — unbranched. So orderly was the chain structure of linear unbranched polyethylene that it emerged from the reaction vessel partially crystallized. Furthermore, if these polymer molecules are deformed either by being stretched or drawn, the chains align themselves along the direction of loading, and pack so well together that they are as much as 80 percent crystalline. The result is high-density polyethylene (HDPE), which has remarkably high strength and stiffness along the chain direction, because the diamondlike carbon-carbon bonds along the backbone must be stretched for their polymer to elongate. Polyethylene and nylon are good examples of polymers that can be strengthened eightfold when they are drawn into fibers. Pulling them through a die mechanically aligns individual molecules by plastic flow, so that the carbon-carbon bond backbone carries the load applied to the fibers.

Many important polymers are closely related to polyethylene in their molecular structures, and they're

233

worth a closer look so that we can appreciate the rela tionship in properties. Hydrogen atoms attached along polyethylene's carbon backbone can be replaced either by single atoms or by groups of atoms, leading to all kinds of new polymers. Taking hints from Ziegler on the role of catalysts in shaping polymer molecules, Giulio Natta, working in the Milan Polytechnic Institute and Montecatini laboratories, managed to synthesize well-ordered molecules of polypropylene, where one of the hydrogen atoms in polyethylene is replaced by a methyl side group. This particular structure is termed isotactic: all of the methyl groups appear on the same side of the polypropylene molecule. These methyl groups introduce steric hindrance in the molecule, making it take on a helical, or helixlike, shape. Its highly ordered structure also means polypropylene is easily crystallized, giving a thermoplastic polymer with superb mechanical properties that can be readily formed into large components, films, and fibers. Polypropylene has a high Young's modulus for a polymer. At room temperature, it can even compete with metals in certain applications, such as in car bumpers. World production of polypropylene jumped from nearly nothing in 1960 to more than 5 million tons annually by 1980. Moreover, polypropylene took only three years to go from discovery to commercialization, as compared to the sixty years it took aluminum, and achieved a similar magnitude of production in only one-quarter the time. It is widely used in automobile dashboards, pipes, and textile fibers. Polypropylene was initially thought of as a replacement for nylon in stockings, except that stockings made from it bag at the knees.

The use of catalysts has accelerated polymeric innovation, making those early synthetic rubbers with branched chains seem like primitive replicas of nature's chemistry. Catalysts developed by scientists at the Goodrich and Firestone research laboratories produced synthetic rubbers with linear chains, close duplicates of nature's handiwork. Working at E. I. Dupont de Nemours in the 1930s, Roy Plunkett was puzzled when he discovered that a cylinder supposedly containing an inert gas, tetrafluoroethylene, seemed to be empty. He sawed the cylinder in half and discovered a solid polymer in which fluorine atoms had replaced all the hydrogen atoms on the polyethylene chain. This remarkable polymer, whose formal name is polytetrafluoroethylene, is trademarked today as Teflon. Teflon is inert, quite resistant to sunlight, moisture, and corrosive chemicals, and has a very small coefficient of friction, making it extremely slippery. The United States government forced Dupont to keep Teflon secret during the Second World War, since it was considered crucial to the war effort in the form of coatings and frictionless surfaces. Most homes today include pots and pans with nonstick Teflon coatings. Anyone who has cooked an omelette in an untreated stainless steel frying pan knows what a boon Teflon is.

Once the idea of replacing some of the hydrogen atoms on the polyethylene chain with other elements was established, a host of new polymers emerged from laboratories. The substitution of chlorine atoms, for cx-ample, produces polyvinyl chloride, also known as PVC. Resistant to chemicals and oils, it is found in containers, pipes, flooring, and raincoats. PVC pipes are gradually replacing metal ones in most home construction. Since the "glass transition" of PVC is above room temperature, its flexibility falls rapidly with a small decrease in ambient temperature, a fact well known to hikers who wear cheap plastic ponchos when the weather turns cold and rainy. Plasticizers (small organic molecules) are added when flexibility is important. Plasticized PVC is frequently used for car upholstery, giving new cars their characteristic smell, which disappears after about one year because of gradual evaporation of the plasticizer. PVC upholstery also becomes brittle with time as the plasticizer disappears — built-in obsolescence.

The addition of large side groups such as benzene rings to polyethylene yields a hard and transparent polymer, polystyrene. When this polymer was first synthesized in the nineteenth century by heating styrene to obtain a solid product called metastyrene, the chemists had no idea what they had done. A lightweight version of polystyrene in the form of foam known as Styrofoam is favored for insulation, protective packing, and the ubiquitous disposable coffee cup. Additions of different large side groups gives us polymethyl methacrylate, known as PMMA and commercially as Perspex, Lucite, and Plexiglas. First developed by the British in the 1930s, the Royal Air Force commandeered the entire output of this tough and transparent polymer for cockpit windows and covers during the Second World War.

This brief overview of a few of the important molecules based on polyethylene illustrates the open-ended nature of polymer-based materials. The possibilities are limitless for adding different side groups to the carbon-backbone chains, creating new forms of molecular architecture. By carrying out reactions between two different kinds of monomers, yet more molecules called "co-polymers" have emerged, some of which are linear, others branched, still others in a three-dimensional network. The Nobel Prize–winning work of Ziegler and Natta established the principles of shaping molecules that crystallize, producing polymers with a relatively high Young's modulus, polymers that today compete with metals where specific properties are important. In the case of car bumpers and side doors, the relatively large elastic deformation associated with polymers is an advantage, since they spring back to their original shape after a minor collision.

The world output of iron and steel has peaked in the past twenty years, but the production of polymers continues to grow. In the last decade, the manufacture of thermoplastic polymers, such as polyethylene, polystyrene, and polypropylene, grew by 60 percent. In the United States alone that translates into an increase in tonnage from fourteen million in 1979 to nearly twenty-two million by 1989. Every few years, either an improved old, or an entirely new polymer emerges from the research laboratories of chemical companies around the world. Looking back over the past thirty years, I would say that Mr. Robinson's advice to Benjamin was excellent.

13

Diamond: The Superlative Substance

W<small>E HUMANS LOVE TO USE SUPERLATIVES.</small> It gives us pleasure to describe someone or something as the "best" or the "fastest" or the "strongest." We have become so inured to this tendency — particularly when practiced by politicians, journalists, or sports broadcasters — that we frequently turn off to what we sense as hyperbole. But now we turn to one of the most intriguing substances found on earth, diamond, whose properties can only be described with superlatives. There is no choice.

Universally considered the most highly prized of all gemstones, diamond's real value to modern industrial societies comes not from its importance to jewelers and their affluent patrons, but from its critical role in more mundane tasks — cutting and grinding. Fortunately for us, industrial diamonds suitable for use as abrasives are one-hundreth the cost of gem-quality stones. A dramatic illustration of the importance of diamonds as cutting tools is in their role in drilling oil wells. Diamond-tipped drill bits last seven times longer than metal bits and can be run twice as fast. Because replacing a drill bit takes oil workers nearly one day, it is estimated that sinking a well to a depth of over 3,300 feet (a typical depth) using diamond-tipped instead of metal-tipped bits saves two weeks and one million dollars.

Carbon can have several different atomic structures. One form we know of as charcoal has an amorphous structure, like glass; another, graphite, the "lead" in pencils, has a hexagonal structure; and finally there is diamond, with its cubic structure. We have already seen how carbon atoms in polymers make four bonds with neighboring atoms, such that the outer tips of the bonds form a tetrahedron. Carbon atoms in diamond are similarly arrayed: one atom at the center of a tetrahedral configuration of four other carbon atoms, building up a three-dimensional and therefore rigid crystal. This three-dimensional network of tetrahedrally arranged covalent bonds gives diamond its remarkable properties.

The word *diamond* may derive from the Greek word for "invincible," *adamas,* because of its great hardness. Indeed, diamond is the hardest material on earth, meaning that it boasts the largest Young's modulus — nearly twice that of any other known substance.[47]

It also has very high thermal conductivity and very low electrical conductivity, behavior wholly different from metals, in which high thermal and high electrical conductivity are typically linked. Aluminum and copper, for example, are used in electrical wiring because of their good electrical conductivity, and in cooking pots because of their good thermal conductivity. Diamond's resistance both to corrosion and to being scratched (except by another diamond) are reflections of its hardness.

Like other highly stiff materials, diamonds are brittle and can be cleaved by a sharp blow parallel to a particular crystal plane. They are also transparent to a wide range of radiation, from visible light, with relatively long wavelengths, to X-rays, with relatively short wavelengths. Finally, when properly cut, diamonds display a brilliance of color that makes them superlative gemstones — with dazzling price tags to match. To the cold dispassionate eye of a scientist, knowing how cheap the other forms of carbon are, this has always seemed preposterous. High-quality blue-white diamonds often sell for $5,000 or more per carat, a term that requires some explanation. Merchants trading in pearls adopted a unit of weight based on the seed of the locust-pod tree, because no matter how old or young the tree, or where it grew, its dried seed always weighed approximately 200 milligrams. This weight was called a "carat," after the Greek *keration,* for locust-pod tree. As might be expected, the exact weight of the locust pod did vary from place to place. In one town it might be 197 milligrams, and in another 207 milligrams. In the early part

Stephen L. Sass

of this century, the carat was standardized to 200 milligrams.

Solid elements frequently exist in two different crystal structures. Iron, for example, can be either BCC (body-centered cubic) or FCC (face-centered cubic). The combination of two external conditions, temperature and pressure, determines which is stable. Once scientists realized that by varying temperature and/or pressure solids could be made to change their atomic structure, they were tantalized by the theoretical possibility that they could manufacture diamonds by putting graphite under high pressures while heating it to high temperatures. The first successful diamond synthesis took place first in Sweden in 1953 and a short time later in the United States. New techniques have even been developed for creating thin diamond films at lower pressures. But before discussing how scientists finally succeeded in transforming graphite into diamonds, let's first trace how we came to know and covet this gemstone whose history is enveloped in romance and intrigue.

Considerable debate still attends the question of when diamonds were first identified. "The sin of Judah is written with a pen of iron, and with the point of a diamond it is engraved on the tablet of their heart, and on the horns of their altar" runs that famous passage in Jeremiah. Even earlier, the Book of Exodus described the garments God chose for Aaron to wear for his consecration into the priesthood:

And you shall make a breastpiece of judgment, in skilled work; like the work of the ephod you shall make it; of gold, blue and purple and scarlet stuff, and fine twined linen shall you make it. It shall be square and double, a span in length and span in breadth. And you shall set in it four rows of stones. A row of sardius, topaz, and carbuncle shall be the first row; and the second row an emerald, a sapphire, and a diamond.

While an engraving tool is certainly a reasonable use for diamond, biblical scholars no longer accept that these quotes refer to what we know today as diamond. Recent translations of Jeremiah replace "diamond" with "adamant," or "hard mineral," and in Exodus with "jade."

An early manuscript recording the economic history of India at the end of the fourth century B.C.E., the *Artha-Sastra* (The Lesson of Profit) of Kautilya, advisor to India's unifier and first emperor Chandragupta Maurya, provides clear evidence that merchants both knew of diamonds and considered them important articles of trade. The *Artha-Sastra* referred to rules for taxing diamonds, called the *Ratnapariska*, evolving over the next millennium into technical manuals known in the West as lapidaries.

In his *Astronomica*, written at the beginning of the first century C.E., the Roman writer Manilius gives the earliest evidence widely accepted as actually describing diamonds in the West. "A small amount of gold," he reports, "exceeds in value countless heaps of brass; the diamond, a stone no bigger than a dot, is more precious than gold." Pliny remarked in his *Natural History* that "the most highly valued of human possessions, let alone gemstones, is the adamas which for long was known only to kings, and to very few of them." And he continues:

> There is the Indian, which is not formed in gold and has a certain affinity with rock-crystal, which it resembles in respect of its transparency and its smooth faces meeting at six corners. It tapers to a point in two opposite directions and is all the more remarkable because it is like two whorls joined together at their broadest parts. It can be as large as a hazelnut. Similar to the Indian, only smaller, is the Arabian.

Pliny informs us that adamas from India came in the form of an octahedron. Octahedral-shaped diamonds are indeed found in nature, so it is believable that the Romans came to know of them as they scoured their empire and its surroundings for valuable resources.

To be brilliant, diamonds must be cut. At the start of the Common Era there was no way to do this, so it is reasonable to wonder why the Romans prized them. Relatively perfect octahedral stones do flash beautiful colors, however, and perhaps it was the unearthing of such rare symmetrical gems that gave diamonds their reputation for possessing magical powers. An Indian manuscript from the sixth century alleges that "he who wears a diamond will see dangers recede from him whether he be threatened by serpents, fire, poison, sickness, thieves, flood, or evil spirits." Romans may have accepted the Eastern notion that adamas conferred invincibility. Such a miraculous property, combined with its rarity and its origin in the distant and mysterious Orient, elevated diamond's value far above that of gold.

Over the centuries bards have woven fantastical tales about diamonds. One such myth involves a so-called Valley of Diamonds, located somewhere in the Scythian desert, between the Black Sea and the Caspian Sea. Epiphanius, a fourth-century bishop, provides a description:

In the Scythian desert, there is a deep valley surrounded by high and rocky mountains. From the summit one cannot see the bottom of the valley, which is lost in the fog as though in the impenetrable depths. The kings of the surrounding countries send their people into the mountains bordering the valley to extract the treasures of precious stones heaped in the farthest depths. But to accomplish this task they must resort to trickery. They kill and flay sheep, then cast the quarters of raw flesh into the depths where the incalculable treasures lie. Soon eagles appear from their aeries; they swoop down through the fog, seize upon the flesh, and carry it back to their nests. The precious stones adhere to this flesh, and king's people have only to rob the eagles' nests to gather them.

Another legend has Alexander the Great using similarly crafty schemes to gather enormous quantities of adamas. Still another involves the adventures of Sinbad the Sailor, as recounted in *The Thousand and One Nights*. After falling into a valley, Sinbad ties a piece of meat studded with diamonds around his waist. A giant bird swoops down and snatches him, lifting him up to its nest.

The names of particular diamonds — the Koh-i-Noor ("mountain of light"), Orlov, Hope, and Regent — evoke regal beauty and mystery. Some were believed to bring misfortune to their owners. The Koh-i-Noor has perhaps the most fanciful and dramatic ancestry. Brahmins in India announced that "he who possesses this diamond, will possess the world. But," they went on to warn, "he will also experience the worst misfortunes, for only a god or a woman can wear it with impunity." Following British annexation of the Punjab in 1849, the Koh-i-Noor was presented to Queen Victoria. She took the old Indian curse to heart and stipulated in her will that should a king inherit the stone only his wife could wear it. In 1937, the Koh-i-Noor was set into the crown of Queen Elizabeth, wife of King George the Sixth. Since then, only the queen mother has worn this crown, which now reposes in the Tower of London.

Before the eighteenth century, all diamonds came from India. The earliest sources are believed to have been the alluvial deposits found along rivers, as well as in pits. Discoveries in Brazil during the early part of the eighteenth century made large quantities of diamonds widely available for the first time. In 1493 Pope Alexander the Sixth divided the Americas between Portugal and Spain along a line of demarcation that brushed the eastern coast of South America, giving Spain everything to its west, including the gold and silver mines that propped up its economy for many years. In 1494, the Treaty of Tordesillas moved this line west, to 48 degrees longitude. The Portuguese king benefited from this west ward movement and quickly took possession of all the diamond-rich territories in eastern Brazil. By 1730, supplies of diamonds in Europe had quadrupled. Early workings were all secondary deposits along rivers, washed down from higher sources. Portuguese prospectors eventually located these primary sites and then had slaves do the work.

Late in the nineteenth century, diamonds were discovered at Kimberley in southern Africa. All earlier sources faded to insignificance in comparison to the vast new mines that were opened. Within a few square miles lie five major sites. The first important one, the Great Hole, is a volcanic pipe containing diamond-bearing minerals — kimberlite — lava pushed up from deep within the earth. Due largely to mines in southern Africa and the synthesis of artificial diamonds, the world's annual output today is approximately 140 million carats. Within a year, 96 percent of these diamonds are consumed by grinding and cutting, leaving five million carats as gemstones.

In our century much attention has turned to the tempting but daunting task of transforming graphite into diamond. The science of thermodynamics, which emerged in the nineteenth century during attempts to improve the efficiency of steam engines, established the principles that guided scientists in their search for a technique to synthesize diamonds.

Thermodynamics seeks to explain the relationship between pressure, temperature, and the stability of carbon's various phases. In particular, it has taught us that graphite is the stable form of carbon at room temperature and atmospheric pressure. Diamond only forms when carbon is raised to high temperatures and high pressures deep within the earth. Upon cooling on the earth's surface, diamond does not revert back to graphite, because its atomic bonds are so strong — and so difficult to break and rearrange at low temperatures. At atmospheric pressure and room temperature, as I mentioned in the discussion on glass, diamond exists as a metastable structure. That is, while it is not the most stable (or lowest-energy) form of carbon under these conditions, diamond remains diamond because the speed with which it turns back to graphite is extremely slow. Experts in the field suggest that the times involved are greater than billions of years. This is lucky for people like Elizabeth Taylor. It might be more than a little disconcerting for her to learn, however, that nature prefers that her magnificent gemstones transform themselves back into pencil lead.

Thermodynamics dictates that other materials — glass and steel, for example — should also decompose. However, while thermodynamics tells us what phase should form during a reaction, it does not tell us whether it will form. The speed, or "kinetics," of the transformation is determined by the difficulty associated with the step(s) in the reaction. The kinetics of the transformation of diamond to graphite is exceedingly slow at room temperature. Diamond is the most spectacular example of the phenomenon of metastability.

Unfortunately for those who have sought to synthesize diamonds, the kinetics of the transformation of graphite to diamond is also very slow. All attempts to obtain diamond by heating graphite to 2000 degrees under a pressure of 50,000 atmospheres — the condition under which diamonds are formed more than 100 miles below the earth's surface — have failed, even though thermodynamics tells us that it ought to form under these conditions. The breakthrough in diamond synthesis came when scientists added iron to the graphite before applying high temperatures and pressures. Iron catalyzed the transformation reaction, allowing millimeter-size diamonds to form. Iron dissolved the graphitic form of carbon into a liquid solution, breaking the strong carbon-carbon bonds, so that upon cooling the carbon atoms were free to rearrange themselves into the diamond structure. Scientists at the General Electric Research Laboratories in Schenectady, New York, synthesized diamonds in late 1954. Immediately after reporting their success in February 1955 they learned to their surprise (and dismay) that two years earlier Baltazar von Platan and his co-workers at the Swedish company ASEA had succeeded in making 0.5-millimeter diamonds, but had not bothered to report their success because they were after gem-quality stones and thought no one else was working on the problem.

The goal had been achieved: graphite was transformed into diamond. It is, however, not possible to synthesize gem-quality stones economically, so the primary uses of synthetic diamond are industrial. Recently, techniques have been developed to deposit thin diamondlike films, which are being pursued because of their great hardness, chemical inertness, and low friction. It would be nice to have large quantities of these films for structural purposes, but this is not yet to be. However, were that to happen, a fundamental question would still be diamond's susceptibility to fracture, which leads us to our next class of materials, designed precisely to overcome this problem.

14

Composites: The Lesson of Nature

ONE OF THE GREATEST CHALLENGES facing materials scientists today is developing substances that can be used under extreme conditions: high temperatures, high stresses, and corrosive environments. Glass and ceramics have the potential of carrying stresses of up to one million pounds per square inch. However, they generally fail under much lower stresses because of the inevitable presence of cracks. The question becomes, how can we take advantage of the high theoretical strength and Young's moduli of such brittle materials while circumventing their potential for catastrophic failure? With glass we saw that the longer the crack the lower the fracture stress necessary to make it fail. Since its discovery, glass by itself has only been used in jewelry, or for containers and windows, and never for load-bearing functions. Ceramics were employed primarily for farm tools, containers, building bricks, tuyeres, and more recently for insulating walls in iron-smelting furnaces. Rarely were they placed in circumstances in which they might fail catastrophically. Consequently bricks used for buildings are stacked on top of one another and always loaded in compression, as is concrete, so that cracks cannot propagate. The first step in putting any brittle material to use in demanding structural situations, therefore, is making certain that they have small dimensions, so that the cracks they contain are all short. For inspiration and guidance, scientists have turned to nature's building materials: bone and wood.

Bone has a complex microstructure. It is composed of thin, hard plates of bone salt, a compound of calcium and phosphate ions embedded in a soft collagen or polymer matrix. Together these constituents make up a "composite" material. We humans have evolved over many millions of years, so it is likely that before bone emerged triumphant from nature's merciless laboratory there were plenty of different structural materials in early life-forms that failed. On a microscopic scale, wood is also a composite of cellulose fibers in a hemicellulose or polymer matrix. On a macroscopic scale, wood is a cellular solid, just like bone, with channels for the passage of liquids. Nature demonstrates that a composite structure, made up of small-diameter fibers (as in wood) or thin, elongated plates (as in bone) embedded in a soft matrix, is superior to a single-phase (one atomic structure) material.

Up to now, the only substances usable in engines and gears have been metals. They can bear heavy loads, endure impacts, and survive in high temperatures without deforming appreciably or fracturing. Metals can also be cast and undergo extensive plastic deformation before fracturing, so artisans could shape them economically and engineers use them with confidence. When strength was called for, metal was the only choice, whether for cannons in the fourteenth century or steam engines in the eighteenth century. Artisans turned first to bronze, then to cast iron, and finally to steel alloys, which have sufficiently high strength and Young's moduli to withstand a wide range of stresses (the latter two were also cheap).

In the twentieth century, however, demand has grown rapidly for substances that can survive in environments where even iron alloys cannot. The jet turbine, for example, requires materials with both high strength and good resistance to oxidation at 1100 degrees or higher, and has spurred the development of an entirely new class of precipitation-hardened nickel alloys. Superalloys, as they're called, have been the mainstay of jet turbine designers for the past several decades. These superalloys contain as many as fourteen different elements — including nickel, aluminum, chromium, ruthenium — but are being applied in circumstances that put them within a mere 75 degrees of their melting point, which leaves little room for improvement. In their search for substances that can survive extremely high temperatures, materials scientists have turned their attention to carbon and boron, and ceramics such as aluminum oxide, zirconium oxide, magnesium oxide, and silicon nitride. A list of melting points can identify the candidates for the next generation of high-temperature materials:

Material	Melting Points, in Celsius
Tungsten	3410
Molybdenum	2610
Niobium	2460
Carbon	3550
Boron	2300
Magnesium oxide	2800
Zirconium oxide	2715
Aluminum oxide	2045
Silicon nitride	1900

Carbon, either in the form of what are called "oriented" graphite plates or in diamond, possesses very high strength. The graphitic form of carbon consists of a stack of planes of hexagonally arranged atoms. The bonding within each plane is strongly directional — "oriented" — or covalent in nature since neighboring atoms share electrons, so consequently its strength and Young's modulus are also high. Weak bonding between the planes of hexagonally arrayed carbon atoms, on the other hand, gives both low strength and low Young's modulus in this direction, which is why graphite is such a good dry lubricant and the graphitic "lead" in a pencil smears easily on paper. However, a fiber of graphite with its planes of strongly bonded carbon atoms oriented along the axis has a Young's modulus twice that of steel, and, because of its low density, a specific modulus more than ten times that of steel. As we've seen, Thomas Edison, in his search for filaments for a lightbulb, had his first success in 1879 when he heated cotton thread in a vacuum, decomposing it to a carbonized fiber.

Unfortunately, graphite fibers are also brittle and cannot be used by themselves in structural applications.

Metals with high melting points, such as tungsten, molybdenum, and niobium, all have body-centered cubic atomic structures and react rapidly with oxygen, and are therefore not good candidates for high-temperature applications. Only metals that form adherent protective oxide films on their surface, such as nickel alloys containing aluminum and intermetallic compounds based on them, can be used safely at elevated temperatures in air. Such coatings could, in principle, also be put on tungsten. But the problem with these films is that a single nick or hole in the film, caused by the impact of, say, a pebble or a grain of sand, would allow the underlying metal to oxidize and lead to the destruction of the metallic component. Should such holes occur in the oxide film on nickel-aluminum alloys, the aluminum in the underlying metal would quickly react with oxygen, reconstituting the aluminum-oxide film, effectively "healing" the minuscule flaw.

Ceramic oxides with melting points above 2000 degrees, the remaining class of materials in the table, are obvious candidates for high-temperature applications because they have already reacted with oxygen and are therefore resistant to further oxidation. The structure of magnesium oxide, where the negative oxygen ions are arranged in a face-centered cubic structure, is similar to that of copper. A magnesium ion sits in the center of an octahedral arrangement of six negative oxygen ions (called an octahedral void). The octahedral configuration first seen in the structure of clay is present here as well, allowing the negatively charged oxygen ions to shield neighboring positively charged magnesium ions from one another. Magnesium oxide crystals are therefore held together by the electrostatic attraction between positively and negatively charged ions. Strong ionic bonds produce stable substances with high melting points.

The zirconium ions in zirconium oxide are also arranged in a face-centered cubic structure. In addition to the octahedral voids I've just mentioned, this structure also has tetrahedral voids, bounded by positive zirconium ions at the corners. If the two negative oxygen ions are placed in the center of all eight tetrahedral voids contained in the unit cell, zirconium oxide is formed. It also has a high melting point. All ionically bonded ceramics can be built up by the periodic stacking of octahedral and/or tetrahedral arrays of positive ions or negative ions. In spite of having a high melting point, zirconium oxide is not by itself a useful structural material at high temperatures, because upon cooling it undergoes a series of changes in atomic structure.

Scientists searching for new high-temperature materials are faced with the critical question of how to use strong but brittle substances safely. The composite structure of bone and wood has provided researchers with hints. A crucial characteristic of the composite structure of bone is its relatively high Young's modulus. Were the bones in our arm to bend when we tried to lift a heavy weight we wouldn't be able to carry out precise manipulations. Bone, in other words, must not deflect appreciably under a load. The Young's modulus of a composite is the sum of the individual moduli of the fiber or plate and the matrix, each weighted by the fraction of the area of each constituent. This is called the "rule of mixtures." Typically, fibers have a high modulus and the matrix has a low modulus. If the composite contains an appreciable amount of fiber, its modulus is determined by the fiber. But given that the fibers are brittle, how can catastrophic failure be avoided? Here is where the soft matrix plays a critical role, though it contributes little to the stiffness of the composite. When a crack propagates across the brittle fiber, it runs into the soft matrix, which absorbs and dissipates much of the energy released by the crack, preventing it from continuing across the composite. Surprisingly, as fibers break, the load-bearing capacity of the composite does not fall rapidly, because there has been a decrease in the area of the strong constituent's cross-section. Rather, the matrix helps to transfer the load from the region of the break to the unbroken portion of the fiber. In a composite made up of strong brittle fibers in a weak ductile matrix, the matrix both redistributes the load away from the break and prevents catastrophic failure.

Possibly the earliest written record of human-made composite structures can be found in the Bible. Exodus recounts how the Hebrews enslaved in Egypt added straw to mud bricks to keep them from cracking while they were hardening in the hot Egyptian sun. The tone of the passage tells us that Pharaoh's heart has also been hardened, and will not be cracked:

Pharaoh commanded the taskmasters of the people and their foremen, You shall no longer give the people straw to make bricks, as heretofore; let them go and gather straw for themselves. But the number of bricks which they made heretofore you shall lay upon them, you shall by no means lessen it; for they are idle; therefore they cry, "Let us go and offer sacrifice to our God." Let the heavier work be laid upon the mean that they may labor at it and pay no heed to lying words.

In this case it is the flexible fibers, the straw, instead of the matrix that prevents fracture. Using the same approach earlier in this century, Bakelite was strengthened and made more resistant to fracture by the addition of small amounts of cellulose from wood pulp. During the dark days of the Second World War, the British, searching for cheap substitutes for aluminum, built Spitfire fighter cockpits out of a composite of flax in a polymer matrix.

Combining dissimilar substances is common practice for materials scientists today, but bow makers in Asia were doing the same thing in a sophisticated manner as early as the third millennium B.C.E. A bow is fundamentally a spring that stores energy when the archer draws the string. Drawing the string places the outside portion of the bow in tension and the inside portion in compression. Any material exposed to high tensile stresses must be able to resist fracture. As it is being drawn, the bow stores the work done by the archer as "elastic energy." Releasing the bowstring accelerates the arrow into flight, transforming the stored elastic energy into kinetic energy. In addition to resistance to breaking, critical issues in the design of bows include the storage of as much energy as possible with the shortest-length bow and the shortest possible draw of the bowstring. Compact weapons are easier to handle — particularly if the archer is mounted on horseback. Artisans needed lightweight materials with high Young's moduli, high yield stress, and high fracture stress, which could also be crafted into the shape of a bow.

The earliest bows were fabricated from a single piece of elm or yew wood, and called "self-bows." To make sure that the stress along the length of the bow was uniform, artisans crafted complex cross-sections and shapes. Bow lengths varied from the six-foot English longbow (so crucial to Edward III's victory at Crécy in 1346), which sought to maximize the velocity and distance of the arrow, to those of the Plains Indians in North America, whose bows had draw lengths of under two feet, convenient for mounted warriors.

Craftspeoples in Asia explored more complicated configurations, taking advantage of the different properties of local materials. Bow makers glued animal sinews to the outside surface of their bows, because sinew has a tensile breaking stress four times that of most woods. This helped to maximize the energy stored in tension. Then artisans realized that they could replace the wood on the inside surface of the bow with a stiffer and stronger substance — animal horn. Their masterpiece was the composite bow, which combined sinews glued to the outer surface of a thin wooden core and horn, typically from water buffalo, glued to the inner surface. Sinew maximizes the tensile stress, while horn, with a strength twice that of hardwoods, carries the compressive stress. Horn also has a higher Young's modulus than wood, which means that more energy could be stored in a composite bow than a wooden bow for the same draw length. The ends of the unstrung bow are reflexed or reversed, so that when the bow is strung, it is under greater tension and more energy can be stored. The composite bow also has a relatively long draw, which means that in combination with its short length it can shoot arrows at a higher velocity and for longer distances than a self-bow of similar dimension. Improvements over the years by the Scythians and then by the Ottoman Turks produced even more powerful bows. The longest shot ever recorded, a distance of 2,915 feet, more than half a mile, was made with a Turkish bow in 1798 by (according to legend) the Ottoman emperor Sultan Selim II. Naturally, the development of gunpowder, cannons, and handguns spelled the end of the bow as a military weapon, and today its use is entirely for sport.

Fiberglass, the first commercially successful composite, consisting of microscopic glass fibers with diameters of 60,000 to 100,000 Å (approximately one-tenth the diameter of human hair) in a polymer matrix, was developed during the Second World War by British and American scientists. Drawn from molten glass that had been passed through hundreds of fine holes in platinum crucibles, these tiny fibers have remarkably high strength, approaching that of steel wire. Fiberglass was initially fabricated for radomes, the covers of airborne radar detectors in Lancaster bombers. A strong lightweight nonmetallic structure was needed, since metals scatter radar waves. Fiberglass went on to have a major impact on the manufacture of fishing poles, building panels, and train and car bodies.

Fiberglass also revolutionized the manufacture of small boats. A mold of the boat hull can be fashioned and then reused many times for layering alternating sheets of polymer resin and glass fibers. In other words, fiberglass allows the mass production of small boats in a process akin to casting, as their hulls are formed in one watertight piece. Also, since the surface of the composite could be made more inert and less absorbent than wood, it was no longer necessary to repaint the hull every few years, easing maintenance. While as strong as steel, glass fibers have a Young's modulus that is much smaller. With 50 percent of its fibers in a weak polymer matrix, fiberglass cannot compete with metals when high stiffness is called for.

Newer composites, reinforced with high-strength and high Young's modulus fibers of boron, graphite, or an oriented polymer such as Kevlar, having similar or lower densities than glass, can replace metal in components where weight is critical. Normally fabricated from long fibers in a polymer matrix, the composite is formed into a tape or sheet, which is wound around an appropriately shaped mandrel or core, and then heated to bond the polymer and fiber together. Careful examination of a modern fishing rod often reveals the spiraling tape along the shaft of the pole. For tennis racquets and fishing poles, where the combination of high stiffness and low weight is desirable, boron and graphite fiber composites have swept away all competing materials.

Compared to fabricating metal, fabricating composites is a complicated process, and consequently costs are high. For example, a one-piece graphite fiber–reinforced composite wing of a vertical takeoff Harrier jet must be cured in an enormous oven during the final step in its fabrication. Aluminum alloys used in aircraft construction cost several dollars per pound, while graphite fiber composites are typically twenty times higher. Sporting goods are often among the earliest uses of expensive new materials, because weekend athletes are quite willing to pay premium prices for the real or imagined advantages bestowed by graphite fiber–reinforced composites, whether in tennis racquets, fishing poles, skis, or golf clubs. Graphite composites are also extremely attractive to the aircraft and aerospace industries, where there is a strong economic incentive to decrease the fuselage weight (and increase the payload). Early nonrecreational uses were in the military and space programs, where high costs are tolerated to a far greater degree than in the private sector. Hence, composites have been fabricated into the movable cargo bay doors and booster rocket casings for the space shuttle, and portions of the fuselage and wings of the latest generation of military jet aircraft. The entire class of Stealth aircraft of the United States Air Force is based on such composites, both because of their impressive mechanical properties and their weak scattering of radar waves. Increasing quantities of composites are finding their way into modern commercial jetliners, such as the Boeing 767 and 777.

Kevlar is a particularly exciting new polymer that can be drawn in such a way as to highly orient its molecules, giving the same advantages as for drawn polyethylene and nylon. A density of 1.45 grams per cubic centimeter — as compared to 2.7 for aluminum, 2.3 for graphite, and 2.5 for glass — and a Young's modulus nearly twice that of aluminum, makes Kevlar an outstanding choice in applications when weight is an issue. A major drawback of all composites based on polymers with carbon backbones is that their wonderful properties degenerate rapidly with increasing temperatures, since carbon reacts strongly with oxygen. Kevlar is superior to most polymers, however, maintaining its high strength up to 180 degrees and high resistance to oxidation until 400 degrees. Kevlar fibers are woven into ropes and cables, and fashioned into composites for bulletproof vests, combat helmets, and lightweight canoes. With a 20 percent annual growth in use during the mid-1980s, Kevlar is clearly a success. More sophisticated formulations evolve with increased usage, and now Kevlar and graphite fibers are being mixed together to make protective helmets with even greater impact resistance.

Because of composites' complex nature, a variety of complications are associated with their fabrication and application. In particular, the strength of a composite material is highly *anisotropic* — that is, strongly dependent on the orientation of the fiber axis — the grain, as it were — with respect to the direction of the load. This behavior is quite different from that of the nearly *isotropic* polycrystalline metals, whose properties are independent of direction. Along the fiber axis, the strength and Young's modulus of composites are quite high, while along the transverse direction (at right angles to the axis, in other words) they are determined by the soft matrix, and accordingly are quite low. Strength and stiffness along the longitudinal and transverse directions can differ by as much as a factor of 1,000, and when engineers design composites they are careful to orient the fibers along the direction of maximum load or weave them together to make the properties more isotropic. Strict quality control of the layering process is also critical; poor bonding between the composite tapes or sheets can be a source of failure. In addition, there must be good bonding between the fiber and the matrix, because if the interface is weak, the fibers will simply pull out, and since they no longer carry any load, failure occurs at stresses far below the design value. For graphite fiber–reinforced polymer composites, where both the fiber and matrix are carbon-based, good interface strength can be achieved because the atomic bonding of both materials is similar, and this leads to strong adhesion.

The need for careful testing is well illustrated by the disastrous experience of Rolls-Royce in the 1970s when they designed a new jet turbine, the RB211, using a graphite-reinforced composite to reduce the weight of the compressor fan blades at the front end of the engine. Unfortunately the designers didn't get around to testing the impact resistance of their compressor blades until quite late in the design process. To the dismay of the engineers, these innovative blades shattered when dead chickens were thrown into the engines (to simulate a bird being sucked into them). Forced into a costly redesign, Rolls-Royce went into bankruptcy and needed a four-billion-dollar injection from the British government to survive. Innovations always carry the risk of failure and the penalties can sometimes be enormous. But so too can the rewards. Composite turbine blades will likely become a reality before the year 2000. Today, composites are being used for parts of outer skin of the turbine intake, where impact resistance is less critical.

For high-temperature applications, thought has been given to combining ceramic fibers and metallic matrices in a composite. Since metals and ceramics have very different types of atomic bonding, their interfacial strength is often quite weak. Improving their performance is the challenge facing the next generation of composites. They are a hot topic for research and the myriad problems associated with bringing together two very different substances, with very different physical and chemical properties, will present a source of both problems and opportunities for some time.

15

The Age of Silicon

To write this book I have been dependent on that marvel of twentieth-century technology, the personal computer. It has completely transformed the workplace, both business and scientific; it dictates the work habits of the student, teacher, scholar, architect, author, and composer; it helps run our government, our military, our media, our lives. In turn, the personal computer is entirely dependent upon the remarkable electrical properties of one material, silicon, and the ingenious methods that have been devised to process it. We have seen that when bonded to four oxygen atoms in a tetrahedral configuration, silicon is an important constituent of clay, ceramic, glass, and the very building material of our earth. It seems nothing less than miraculous that elemental silicon can also be transformed into the substance of the brains and memories of our computers.

A computer is a device for storing and processing information rapidly, following a set of instructions called a program. It does all this electronically. In the nineteenth century, calculating engines, the ancestors of

265

computers, were entirely mechanical. To come up with devices to compute, it was necessary to devise a system that could store both the facts (either numbers or words) and the instructions on how to manipulate them (the program). How can a computing engine store numbers electronically? For everyday arithmetic we use the decimal or base-ten system, the Arabic notation we discussed earlier, in which any number can be represented using ten different digits: 0,1,2,3,4,5,6,7,8,9. Take the number 328. Remember each place position has a value given by 10 raised to a power n, that is, 10^n. Going right to left, in the first position, the ones' place, n is 0; in the second position, the tens' place, n is 1; and so on. For the number 328, the first digit, 8, is multiplied by 10^0, which equals 1, the second digit, 2, is multiplied by 10^1, which equals 10, and the third digit, 3, is multiplied by 10^2, which equals 100. The number 328, then, is the sum of 8 plus 20 plus 300. In order to use the base-ten system in the computer we would have to find a way to represent ten different digits at each place position. This is complicated. Better would be a numeration system in which there are only two choices for each place position. This would allow use of some kind of switch to represent numbers, like a light switch that is either in the "on" or "off" positions. The binary or base-two system of representing numbers permits this.

In the binary system only two digits are used, 0 and 1. So, how to represent a number in this system? Consider the number 111_{two}, with the subscript "two" reminding us that we are using the base-two system. Remembering how we assigned a value to each place in the base-ten system, starting on the right, in the base two system the first 1 is multiplied by 2^0, which equals 1. Moving left to the next position, the second 1 is multiplied by 2^1, equaling 2, and finally the third 1 is multiplied by 2^2, which equals 4. To get the number 111_{two} in the base-ten system, we add $1(2^0)$ plus $1(2^1)$ plus $1(2^2)$, which equals 1 plus 2 plus 4 and obtain 7_{10}, where the subscript 10 (normally not used) reminds us we are in the base-ten system. In other words, the number 111 in the base-two system equals seven in the base-ten system. How do we represent our previous example, 328, in the binary system? Using the approach just described, we can write 328_{ten} as 101001000_{two}. This can be converted back to base ten as follows: 101001000_{two} equals: 1 times 2^8, plus 0 times 2^7, plus 1 times 2^6, plus 0 times 2^5, plus 0 times 2^4, plus 1 times 2^3, plus 0 times 2^2, plus 0 times 2^1, plus 0 times 2^0. And that equals 328_{ten}. Clearly, the base ten, or Arabic, notation is a far more compact and efficient way of representing numbers. Yet however clumsy it is, the binary system is now universally being used to represent numbers in a computer. And this is because of the electronic switches.

If we can imagine a light switch for each place, the number 111_{two} can be represented by three switches in a row in the "on" position, each passing electric current. With the help of a device for storing an electric charge, such as a capacitor, the switch can be used to charge the capacitor, which then would represent 1. If the capacitor was discharged, 0 would be represented. If, for example, we flipped off the rightmost switch and drained the capacitor, we would have 110_{two}, which is 6 in the base-ten system. The simplest way to represent numbers electronically is with a switch and a way to store and detect an electric current. While easy to visualize, a light switch is not very practical for our purposes, since it is both large and must be thrown mechanically. Switches of that nature are similar to those actually used in the earliest computers, such as the Electronic Numerical Integrator and Computer (ENIAC), built in the Moore School of Electrical Engineering at the University of Pennsylvania at the end of the Second World War. The ENIAC's entire memory consisted of twenty ten-digit decimal numbers, requiring 6,000 switches covering three walls, which took days to reprogram. A marvel of its time, the ENIAC could perform 333 multiplications per second. But another device that can be made to either pass or not pass current, and has the potential to be made with extremely small dimensions, is the transistor, based on a class of materials called semiconductors.

I noted earlier that metals are excellent electrical conductors. When we apply a voltage from a battery to a metal wire, an electrical current flows easily for even the smallest applied voltage. The equation governing the current, denoted I, is Ohm's law, which states that E equals I times R, where E is the applied voltage and R is the resistance of the material to the passage of current. Ohm's law tells us that as the voltage increases, the current also increases, and in a linear manner. Metals have large numbers of free electrons that are not bound to individual atoms in the crystals, and these negatively charged particles move easily under the force of an applied voltage. The actual current we get is determined by the resistance. For a fixed voltage, the current becomes smaller as the resistance becomes larger.

Compare this behavior to that of a semiconductor, such as silicon, when the voltage is increased. For small voltages there is no current. Only when the voltage reaches a threshhold value, in this case 1.1 volts, does an appreciable current begin to flow, and then there is a rapid increase before leveling off. This strong nonlinear response of the current versus the voltage is a unique characteristic of a semiconductor, and allows it to act like a switch: the current is turned on and off by the applied voltage. Before seeing how this behavior forms the basis of a storage device in a computer, let's look at its origins.

This is best done by comparing the properties of aluminum, which exhibits typical linear metallic behavior, and silicon, which as we've just seen exhibits nonlinear semiconductor behavior.[48] Aluminum and silicon are neighbors on the periodic table of the elements. Aluminum has thirteen protons and thirteen electrons, and silicon has fourteen protons and fourteen electrons; so the atomic numbers of aluminum and silicon are thirteen and fourteen, respectively. If you examined them, they would look similar, each metallic in appearance, one silvery, the other blackish. They both feel light, because they have similar densities — aluminum, 2.7 grams per cubic centimeter, and silicon, 2.3 grams per cubic centimeter. How do they differ? Measuring their electrical resistivity (the R in Ohm's law), you find that aluminum conducts electrical currents 10^{11}, or eleven orders of magnitude, better than silicon. This difference is related to the quite different voltage-current behavior we just discussed. Also, aluminum becomes molten at 660 degrees, and silicon at 1414 degrees. The difference between their melting points tells us about the bonding between their atoms. It is clearly much stronger in silicon than in aluminum. Further, try and bend a one-inch diameter bar of each metal across your knee and you will discover that pure aluminum plastically deforms easily, while silicon breaks like glass. Again, this is because the bonding between the atoms is quite different.

A good way to visualize bonding is to imagine forming a solid by bringing together distant atoms and examining what happens to their energy as their separation decreases. Atoms feel forces between them that, if attractive, can be thought of as bonds. In aluminum a cloud of free electrons surrounds the periodic array of atoms. These electrons move easily under an applied voltage, which explains why metals conduct electrical currents easily. We saw earlier that silicon likes to bond to four other atoms, which are arranged on the corners of a tetrahedron. When electrons are not free to wander about, as in a metal, but are instead constrained to be shared between two atoms, the bonding is covalent. An appreciable amount of energy is required to break an electron free from its special location between atoms; for silicon, as we've seen, this energy is 1.1 volts. Once the covalent bond is broken, the electron is free to move through the crystal. And so we have our light switch.

The actual process of creating this silicon switch is more involved than this. For example, for the silicon switch to perform properly, it is necessary to add small quantities of "dopants" to form a "field effect transistor," which was invented by William Shockley, in collaboration with John Bardeen and Walter Brattain, at Bell Laboratories. All three shared a Nobel Prize for their work. The field effect transistor is like a triode vacuum tube, which acts as an amplifier. A voltage — let's say 1.1 volts, which is sufficient to cause a current to flow — is applied between the source and drain contacts. Whether a current actually flows is decided by small changes of the voltage on the gate between the source and the drain. A voltage of −0.1 volts there causes the actual voltage between source and drain to be 1.1 volt minus 0.1 volt, or 1.0 volt, which is insufficient to cause an electron to break free from the covalent bond. Because small changes in the voltage on the gate cause large changes in the current flowing from the source to the drain, the field effect transistor behaves like an amplifier: it takes a weak signal and magnifies it. This is the same behavior exhibited by a triode vacuum tube. Dopants are introduced to control the conductivity of silicon and to change the energy required to move an electron from being nonconducting to conducting. This behavior underlies the operation of all computers, because transistors can now be used to store information in the binary system of numeration.

Transistors were an enormous advance over vacuum tubes, which are fragile, bulky, require enormous amounts of cooling, and have a relatively high failure rate. Early designers built computers by assembling small field effect transistors individually, and then mounting them one by one on a board. Making and mounting individual transistors was a clumsy, difficult task, and since a single transistor-capacitor combination was required to store the contents of one place in binary notation — one "bit" of information — many transistors were required to store an appreciable amount of information.

In 1957, Fairchild Semiconductors in Mountain View, California, was in the business of manufacturing silicon transistors. Their workers first made a number of transistors on a thin, single-crystal silicon wafer, then cut them out individually, and finally wired them together to make an electronic circuit. Early transistor radios and hearing aids were assembled in this manner. In addition to the time-consuming and tedious nature of the wiring technique — as well as the unavoidable errors in manufacturing — the speed of the computer was limited by the length of the connecting wire, even though the electronic pulses themselves traveled at the speed of light. As more complex circuits were built and the length of wiring increased, so did the time required for the calculation. To speed things up, it was necessary to make the path lengths shorter, which meant making the transistors smaller, packing them more closely together.

The question became how to do this. Silicon crystals used for the fabrication of electronic devices needed to be perfect, because defects change the local electrical behavior of a semiconductor. This means no dislocations and grain boundaries can be present. Remarkably, silicon can now be grown in the form of perfect single crystals twelve inches in diameter and several feet long. These crystals are then carefully sliced into thin wafers, like a salami. Even dust particles on the surface can ruin the electrical behavior of silicon. In 1958, Jean Hoerni, who worked at Fairchild, developed a technique for protecting the transistor chips from dust particles by depositing a thin layer of silicon dioxide in a planar geometry on the surface.

While pondering the problem of simplifying the manufacture of transistors, Robert Noyce, who also worked at Fairchild, considered this planar process of forming silicon dioxide and realized that it might be possible to form an entire electronic circuit — transistors, resistors, and capacitors — on the same silicon chip, using a similar deposition technique. The silicon dioxide acted as an insulator to isolate each component. Noyce found that adding dopants to special sites allowed the fabrication of individual electronic components; one dopant produced a transistor and another a resistor. Noyce then tackled the challenge of linking together the different elements by metallic conductors. Instead of connecting them with thin wires, he came up with the idea of printing metal lines directly on the surface of the silicon dioxide–coated chip. This allowed separate regions of an individual transistor to be joined, and then these metallic lines could be used to link transistors to each other. The integrated circuit, the heart of all modern computers, was born.

The market for these devices was first driven by the Minuteman missile program, and later the Apollo space program, which sought to meet John F. Kennedy's challenge to put a man on the moon by the end of the 1960s. Miniaturized on-board computers were needed to control both the rockets and the lunar modules. Therefore, most of the first sales of these early chips were to the military and then the space programs, illustrating how defense-related needs often drive, or at least pay for, revolutionary advances in technology. In 1964, five years after Noyce invented the microchip, ten circuits were being put on the same chip. By 1969, 1,000 circuits were being placed on one chip; by 1975, 32,000 circuits, creating the 32 Kb memory chip. Today several million transistors can be fabricated on a single chip. The goal is to put a billion transistors on a chip by the year 2000. The name of the game is making the transistors as small as possible and the paths between them as short as possible, since the number of calculations that can be performed in a second is controlled by the time required to send electronic pulses between different components of the circuit. Current technology makes transistors that have dimensions on the order of a micrometer or less and metal lines with widths of 2,500 Å (0.25 micrometers), dimensions which can be compared to the fifty-micrometer diameter of human hair. These widths will diminish by half or more in the very near future.

Noyce left Fairchild and went on to help start a new company called Intel, which today dominates the marketplace for microprocessors, the brains of all computers. Intel's first products were the 4004 and 8008 memory chips, both based on the integrated circuit. At the same time that Noyce was coming up with the concept of putting electronic elements on a silicon chip, Jack Kilby was pursuing the same goal at Texas Instruments. Kilby was the first engineer to build an integrated circuit on a chip, which he demonstrated to the executives at Texas Instruments in 1958. Unfortunately, he did not connect the separate electronic elements with evaporated metal lines deposited on the surface of the chip, so the U.S. Patent Office gave precedence to Noyce, because he was the first to solve the interconnection problem. Today, however, Robert Noyce and Jack Kilby are considered the co-inventors of the integrated circuit. Based on Kilby's ingenuity, Texas Instruments turned out the first practical hand-held pocket calculator. Initially selling for several hundred dollars, simplified versions are now on sale at supermarket counters everywhere for under ten dollars. Calculations that used to be done by slide rule can now be performed with calculators the size of credit cards. A curious and worrisome outcome is that the calculations science students once performed in their heads they no longer can. Like others, I seem to have lost the facility to manipulate numbers. I sometimes wonder if the minds of the next generations of engineers and scientists will be rendered too lazy by these electronic slide rules. We have gained a great deal by these marvels. What have we lost?

In the 1960s another Intel employee, Ted Hoff, came up with a revolutionary idea: instead of designing special components for particular calculations in, for example, a pocket calculator, why not put all the functions of a computer — the logic circuits, arithmetic calculation circuits, and printer commands — on a single silicon chip, with separate chips holding the input-output and program components? Hoff's first microprocessor, which contained more than 2,000 transistors on a chip measuring one-eighth of an inch by one-sixth of an inch, was as powerful as those early computers that filled air-conditioned rooms in the 1950s and 1960s. His invention started a revolution, since these tiny chips could be produced very cheaply — for about $100 in 1971, for five dollars by the late 1970s. Miniature computers could now be put inside a refrigerator, or an automotive electronic ignition system, or a microwave oven for precision control. The Intel 8080 microprocessor chip, the first in a long series, was capable of executing 290,000 operations per second. The latest in the Intel series, the Pentium, executes 100,000,000 operations per second. In the 1940s the ENIAC, which performed 333 multiplications per second, needed more than 17,000 vacuum tubes and 6,000 switches, and was an eighty-foot-long, thirty-ton giant. In nearly fifty years that performance has improved by a factor of approximately one million, thanks to a silicon chip the size of my fingernail, inside a small box weighing a few pounds.

The size and spacing of the transistors on the silicon chip, the widths of the metal lines joining them today, the thickness of various layers on the chip, are all well below 10,000 Å. Some dimensions go down to 100 Å or less. Scientists at companies like Intel, Motorola, and IBM will continue to shrink these dimensions. At some point, however, the miniaturization process must run into problems, because it is apparently not possible to go below the size of individual atoms. What then? Scientists on the laboratory bench are now able to manipulate individual atoms. Will they be able to build devices based on a single atom? There lies the challenge for the years ahead.

Epilogue

Materials in the Twenty-First Century

> My aim is the same as in locomotive building: to arrive at a given functional form in the most efficient manner.
>
> — David Smith, sculptor, 1952

IT WAS ANOTHER of those impossibly hot and humid summer afternoons in Ithaca. I wandered restlessly around the house, doing small chores to pass the time. Standing over a pyramid of flowerpots on our patio, I watered each desiccated, drooping plant in turn. Looking around my yard, I idly reviewed the materials that shape my life and make it so comfortable. A rubber hose, clay flowerpots neatly arrayed on stone-stacked terraces, my all-metal (except for the rubber tires) garden tractor, a wood-frame ranch-style house — these materials have been known to humans for hundreds, if not thousands of years. But inside our house are a CD player, several computers, and a television set — all based on a silicon technology that has emerged in the past forty years.

I thought back to my boyhood in New York City in the early 1950s. My greatest pride and joy then was my Hallicrafters S38-C shortwave radio receiver, which I bought (for less than $100) with money I had earned from working as a delivery boy for my father. I earned 75¢ an hour, plus 25¢ tips for carrying sacks of meat up and down the hilly tree-lined streets of the west Bronx. Into the early morning hours I would sit hunched over my Hallicrafters, lovingly tuning the dials, using the band spread to extract weak radio signals from the other side of the world, from places I could only dream of. Today, for only a little bit more money, you can buy a far better shortwave receiver that fits into the palm of your hand, one that uses an antenna only a few inches long instead of the fifty feet of copper wire I had to string across the roof of my six-story apartment house. But why even bother with shortwave today, when we can so easily receive sharp TV images and retrieve information using the Internet from anywhere in the world? As a young student I had not the slightest idea what kind of world awaited me. I never imagined that within only a decade the bulky vacuum tube amplifiers of my radio would be replaced by tiny transistors on silicon chips. Did anyone?

Information and communication technology, transportation, recreation, human orthopedic implants — these are all areas already transformed by a revolution

Stephen L. Sass

in materials: silicon transistors replacing vacuum tubes; glass fibers supplanting copper wires; jet turbines displacing piston engines; graphite fiber–reinforced tennis racquets and fishing poles superior to those made from wood; ceramic and metal hip joints replacing nature's own. Can the pace of these changes continue unabated into the twenty-first century? This is a question technological prophets attempt to answer at the peril of their reputations.

Certainly things will change. Let's start with what we know we would like to do better, even though some might question whether our lives (or the earth) will be the better for it. High-speed travel from New York City to Tokyo would be a wonderful boon to business people. This is, or was, the goal of the National Aerospace Plane (NASP), ambitiously put forward by NASA in the 1980s. It was predicted that the NASP would take off from Kennedy Airport in New York, go into space, fly at twenty-five times the speed of sound, and then glide into Narita Airport in Tokyo three and one-half hours later. The project has now been quietly shelved; there are no materials available for either its fuselage or its engines. A more modest proposal is the High Speed Civil Transport, designed to carry passengers at two to three times the speed of sound — a sort of economical Concorde. But the engines that will power it do not yet exist. To be more efficient than today's engines, and to leave no nitrous oxide pollutants in the ozone layer, which is where the next generation of aircraft will fly, they must reach temperatures of 1400 degrees, 200 to 300 degrees higher than is possible today. At these high temperatures it is unlikely that metallic alloys alone can be used successfully, so engine designers must turn to ceramics, which are attractive because of their high melting temperatures and good oxidation resistance, worrisome because they fail catastrophically on impact.

What can be done? As I suggested at the end of chapter 14, nature has provided the outline of a solution; that is, a composite material that combines ceramic with a constituent that will prevent cracks from propagating across, for example, a turbine blade. Of course, both components must be able to withstand hot burning gases. The choices for the ductilizing constituent are limited: metal, ceramic, or an intermetallic compound. Since some ductility is needed, the most obvious choice is a metal. By itself metal would be too weak at these elevated temperatures, but it might endow a ceramic with the necessary forgiveness, allowing it to undergo what we call "graceful failure," meaning failure that gives plenty of warning before disaster occurs. There are currently no metal-ceramic composites able to survive these elevated temperatures. Some scientists have decided that metals will never be used successfully in these circumstances and instead have focused on composites made up entirely of ceramics. One promising candidate is silicon nitride, which can be used to reinforce itself as a strong ceramic, with turbine parts fabricated from a ceramic-ceramic composite. Though its fracture resistance is much greater than unreinforced silicon nitride, it is still not nearly as resistant to fracture as a metal. The search will likely continue well into the next century.

Tripling the fuel efficiency of cars will be a primary

goal in the twenty-first century. We are a long way off. Detroit's wonder car will have to be able to get between seventy and eighty miles per gallon, and today's average is between twenty-five to thirty miles per gallon. What is necessary is decreasing the weight of the car, following the path pioneered by the aircraft industry, where the name of the game is thrust-to-weight ratio — useful work output to weight of the engine. This is why modern aircraft use more and more carbon-based composites in their fuselages. Also, as the operating temperature of the aircraft engine increases, so too does its efficiency and, therefore, its thrust-weight ratio. What can Ford, General Motors, Chrysler, Honda, and Toyota do to make their automobiles lighter? Begin incorporating more polymers or polymer composites into the structural components of their cars, since polymers, made up largely of carbon and hydrogen — both light elements — are remarkably lightweight. And make the engine of ceramics, which, in addition to their light weight, require less cooling than cast-iron or aluminum engines.

As we saw in the last chapter, the modern computer is the result of the large-scale miniaturization of electronic components. Building on the capabilities to make things smaller and smaller, interest increases in making mechanical devices such as motors and sensors on a tiny scale. Just recently a colleague of mine at Cornell produced a silicon guitar several microns in length to demonstrate what is now possible with respect to miniaturization. Perhaps the future will see little mechanical devices at nearly atomic proportions doing our bidding. These exist only in the imagination of materials scientists and mechanical engineers. But that they exist there means that one day they may be realized.

Are there other, wholly different solutions to such challenges that ingenious scientists will devise or, as is even more likely, stumble across? That possibility keeps scientists and engineers laboring in their laboratories day and night, seven days a week. Perhaps Adam Smith was right: those driven purely by curiosity and by a willingness to ask seemingly irrelevant questions will discover the substances that shape the course of history. We saw how the printing process benefited from capabilities developed by Flemish artists; perhaps new visions will emerge from a union of art, engineering, and science, all of which, as the American sculptor David Smith reminds us, share common goals. Whoever it is must have the ability to combine, as Adam Smith put it, "the powers of the most distant and dissimilar objects." Upon that ability human civilization has always depended.

Notes

1. J. S. Cooper and W. Heimpel, "The Sumerian Legend," *Journal of the American Oriental Society* 103 (1983): 67–82.

2. H. W. F. Saggs, *Everyday Life in Babylonia and Assyria* (New York: G. P. Putnam and Sons, 1965) p. 32.

3. The total number of unit cells in this crystal is huge, and best expressed by the use of scientific notation; that is, by the powers of ten, where $10 = 10^1$; $100 = 10^1 \times 10^1 = 10^2$; $1,000 = 10^1 \times 10^1 \times 10^1 = 10^3$; $1,000,000 = 10^1 \times 10^1 \times 10^1 \times 10^1 \times 10^1 \times 10^1 = 10^6$. Note that when you multiply numbers having the same base (in this case 10) you add the exponents. In this notation 25 million is written as 25×10^6 and the number of unit cells in the cube-shaped crystal is calculated as 25 million x 25 million x 25 million $= (25 \times 10^6) \times (25 \times 10^6) \times (25 \times 10^6) = 25 \times 25 \times 25 \times 10^6 \times 10^6 \times 10^6 = 15,625 \times 10^{18}$ or 1.5625×10^{22}. A very large number indeed! For comparison, the United States debt in 1995 was 5 trillion dollars; that is, \$5,000,000,000,000 or 5×10^{12} dollars.

4. G. Agricola, *De Re Metallica,* trans. by H. C. and L. H. Hoover (reprint, New York: Dover Editions, 1950) p. 34.

5. J. B. Pritchard, ed., *Ancient Near Eastern Texts Relating to the Old Testament* (Princeton: Princeton University Press, 1955) pp. 229–30.

6. L. Legrain, "Business Documents of the Third Dynasty of Ur," *Ur Excavation Texts* 3, no. 385 (1937).

7. T. A. Wertime and J. D. Muhly, *The Coming of the Age of Iron* (New Haven: Yale University Press, 1980) p. 46.

8. P. R. S. Moorey, "Materials and Manufacture in Mesopotamia: The Evidence of Archaeology and Art, Metals and Metalwork, Glazed Materials and Glass," *British Archaeological Reports,* 237 (1985): 44.

9. B. Kienast, *Die Altassyrischen Texte des Orientalischen Seminars der Universitat Heidelberg und der Sammlung Erlenmyer-Basel.* Reproduced in K. R. Veenhof, *Aspects of Old Assyrian Trade and Its Terminology* (Leiden, Holland: 1972) pp. 307–8.

10. For purists, the phases of pure iron are slightly more complex than described here. For our purposes, there is no need to go into this complication.

11. A. Goetze, "Kizzuwatna and the Problem of Hittite Geography." (New Haven, Yale University Press, 1940) pp. 29–39.

12. J. Chadwick, *The Mycenaean World* (Cambridge: Cambridge University Press, 1976) p. 141. See also Wertime and Muhly, *Coming of the Age of Iron* p. 44.

13. Wertime and Muhly, *Coming of the Age of Iron,* p. 44.

14. George Herbert Palmer, trans., *The Odyssey of Homer,* bk 2 (Boston: Houghton Mifflin, 1886) p. 305.

15. *British Archaeological Reports,* vol. 237 (1985), p. 103.

16. R. Pleiner and J. K. Bjorkman, "The Assyrian Iron Age," *American Philological Society Journal,* 118 (1974): 283.

17. Anita Engle, *Readings in Glass History,* vol. 1 (Jerusalem: Phoenix Publications, 1973) p. 2.

18. Chloe Zernick, *Short History of Glass* (Corning, N.Y.: Corning Museum of Glass Press, 1980) p. 17.

19. Engle, *Readings in Glass History,* pp. 36–37.

20. A. L. Oppenheim, R. H. Brill, D. Baraga, and A. Von Saldern, *Glass and Glassmaking in Ancient Mesopotamia,* (Corning, N.Y.: Corning Museum of Glass Press, 1988) p. 73.

Notes

21. Zernick, *Short History of Glass,* pp. 21–22.
22. N. K. Sandars, *The Epic of Gilgamesh,* prologue (New York: Penguin Classics, 1971) p. 59.
23. Vitruvius, quoted in R. H. Bogue, *Chemistry of Portland Cement* (New York: Reinholt Publishing, 1947) p. 5.
24. Pliny the Elder, *Natural History,* quoted in R. Parkinson and S. Quirke, *Papyrus,* (Austin: University of Texas Press, 1995) p. 9.
25. See C. J. Humphreys and W. G. Waddington, "Dating the Crucifixion," *Nature,* 306 (1983): pp. 743–46.
26. Zernick, *Short History of Glass,* p. 42.
27. J. G. Thompson, *Mining and Metallurgy,* (New York: American Institute of Mining and Metallurgical Engineers publication, May 1940).
28. Wang Chen, *Treatise on Agriculture (Nung Shu),* cited in Joseph Needham, *Science and Civilisation,* vol. 5, pt. 1 (Cambridge: Cambridge University Press, 1985), pp. 206–207.
29. Roger Bacon, *Thirteenth-Century Prophecies,* quoted in *A History of Technology,* vol. 3, ed. C. Singer, E. J. Holmyard, A. R. Hall, and T. I. Williams (Oxford: Oxford University Press, 1957) p. 719.
30. Ibid., p. 650.
31. Lacey Baldwin Smith, *The Horizon Book of the Elizabethan World* (Boston: Houghton Mifflin, 1967) p. 22.
32. Needham, *Science and Civilisation in China,* vol. 5, p. 112.
33. Ibid., pp. 118–122.
34. Ibid., p. 167.
35. A. Y. al-Hassan and D. R. Hill, *Islamic Technology* (Cambridge: Cambridge University Press/Unesco, 1986) p. 112.
36. Quoted in D. A. Fisher, *The Epic of Steel* (New York: Harper & Row, 1963) p. 31.

37. Quoted in *Economic History of Europe,* vol. 4, ed. E. E. Rich and C. H. Wilson (Cambridge: Cambridge University Press, 1967) p. 113.

38. Adam Smith, quoted in *A History of Technology,* vol. 4, ed. Singer et al., p. 150.

39. The principle of the conservation of matter holds, except if individual atoms split apart or fuse together, which can happen under special and extreme conditions when matter can be destroyed by conversion into energy. This is the basis of nuclear weapons, which are classified as either fission or fusion devices, depending on whether atoms are split or fused.

40. Why a crack must achieve a critical length before growing rapidly is complicated to explain. Briefly, when a solid is elastically stressed it stores energy in the form of stretched atomic bonds. As a crack grows it releases some of this stored energy, which can contribute to the crack increasing in length. At the same time the crack grows, its surface area also increases, and this new area has extra energy associated with it, since the atomic bonding across the surface is different (both in number and strength of bonds) from inside the perfect crystal. A crack achieves its critical length when the energy released by its growth is just enough to create the two new surfaces, and then it is poised to grow catastrophically. This explanation holds best for brittle substances like glass and ceramics, and is less valid for ductile materials such as metals, where the high stresses at the tip of the crack can be relieved by plastic deformation processes involving the motion of dislocations.

41. A lens has a symmetry line normal to itself called an optic axis. The focal length is the distance along the optic axis from the lens to the point where rays coming from a distant object (e.g., the sun) are focused. This point is called the focal

point of the lens. The maximum temperature is at the focal
point.

42. The reacting molecules are believed to align themselves
on the periodic atomic structure of the surface, which acts as
a template, so that different molecules are brought close together in a manner that speeds their interaction.

43. Henry Bessemer, *Sir Henry Bessemer: An Autobiography*
(London: n.p., 1905) p. 170.

44. Charles Goodyear, *Gum Elastic and Its Variations, With
a Detailed Account of Its Applications and Use*, vol. 1 (New
Haven, n.p., 1855) p. 23.

45. R. Hooke, *Micrographia* (New York and London: Dover
Editions, 1961) p. 7.

46. This is one of two common polymerization processes,
with the second being addition, whereby a single simple molecule, known as a monomer, is added to a string of others to
build up the long polymer chain. One form of synthetic rubber is produced by a condensation reaction using an isoprene
monomer.

47. There is one exception to this statement, since boron nitride has a stiffness approximately three-quarters that of diamond.

48. See J. W. Mayer and S. S. Lau, *Electronic Materials Science: For Integrated Circuits in Si and GaAs* (New York:
Macmillan, 1990).

Bibliography

Adams, Robert. *Heartland of Cities.* Chicago: University of Chicago Press, 1981.

Agricola, G. *De Re Metallica.* Translated from the first Latin edition of 1556 by H. C. Hoover and L. H. Hoover. Reprint, London: Dover Editions, 1950.

al-Hassan, A. Y., and D. R. Hill. *Islamic Technology.* Cambridge: Cambridge University Press, 1986.

Bessemer, Henry. *Sir Henry Bessemer: An Autobiography.* London, n.p., 1905.

Bogue, R. H. *Chemistry of Portland Cement.* New York: Reinholt Publishing, 1947.

Chadwick, J. *The Mycenaean World.* Cambridge: Cambridge University Press, 1976.

Ciardelli, F., and P. Giusta, eds. *Structural Order in Polymers.* Florence, Italy: Lectures presented at the International Symposium on Macromolecules, September 1980. Oxford: Pergamon Press, 1981.

Cottrell, A., ed. *Ancient Civilizations.* London: Penguin Books, 1988.

de Jesus, P. S. "The Development of Prehistoric Mining and Metallurgy in Anatolia," *British Archaeological Reports* 74 (1980).

Engle, Anita. *Readings in Glass History,* vol. 1. Jerusalem: Phoenix Publications, 1973.

Fisher, D. A. *The Epic of Steel.* New York: Harper & Row, 1963.

Garnsey, P., and R. Saller. *The Roman Empire.* London: Duckworth Press, 1987.

Goetze, A. "Kizzuwatna and the Problem of Hittite Geography." New Haven: Yale University Press, 1940.

Goodyear, Charles. *Gum-Elastic and Its Variations, With a Detailed Account of Its Applications and Use, and the Discovery of Vulcanisation,* vol. 1. New Haven: n.p., 1855.

Gordon, J. E. *The New Science of Strong Materials.* 2d ed. Princeton, N.J.: Princeton Science Library, 1976.

Hooke, R. *Micrographia.* London: Publication of The Royal Society, 1665. Reprint, New York: Dover Editions, 1961.

Kampfer, F., and K. G. Beyer. *Glass: A World History.* New York: New York Graphic Society, 1966.

Kelleher, Bradford D. *et al. Treasures of the Holy Land.* New York: Metropolitan Museum of Art, 1986.

Kempinski, A., and S. Kosak. "Hittite Metal Inventories." Tel Aviv: *Tel Aviv,* vol. 4, 1977.

Kenyon, Kathleen. *Archaeology in the Holy Land.* New York: Praeger Press, 1966.

Lea, F. M. *The Chemistry of Cement and Concrete.* 3d ed. New York: Chemical Publishing, 1970.

Maillard, R., ed. *Diamonds.* New York: Crown Publishers, 1980.

McEwen, E., R. L. Miller, and C. A. Bergman. "Early Bow Design and Construction." *Scientific American,* June 1991.

McMillan, F. M. *The Chain Straighteners.* New York: Macmillan, 1979.

Mellaart, James. *Earliest Civilizations of the Near East.* New York: McGraw-Hill, 1965.

Mellaart, James. *Çatal Hüyük.* London: Thames & Hudson, 1967.

Moorey, P. R. S. "Materials and Manufacture in Mesopotamia: The Evidence of Archaeology and Art, Metals and Metalwork, Glazed Materials and Glass." *British Archaeological Reports,* 237 (1985).

Muscarella, Oscar White. *Bronze and Iron*. New York: Metropolitan Museum of Art, 1988.

Needham, Joseph. *Science and Civilisation in China*, vol. 5. Cambridge: Cambridge University Press, 1985.

Oppenheim, A. Leo. "Mesopotamia in the Early History of Alchemy." *Revue d'Assyriologie* 60 (1966).

Oppenheim, A. L., R. H. Brill, D. Barag, and A. Von Saldern. *Glass and Glassmaking in Ancient Mesopotamia*. Corning, N.Y.: Corning Museum of Glass Press, 1988.

Palmer, George Herbert, trans., *The Odyssey of Homer, bks 1-12*. Boston: Houghton Mifflin, 1886.

Parkinson, R., and S. Quirke. *Papyrus*. Austin: University of Texas Press, 1995.

Pleiner, R., and J. K. Bjorkman. "The Assyrian Iron Age." *American Philological Society Journal*, 118 (1974).

Postgate, J. N. *Early Mesopotamia*. London and New York: Routledge, 1994.

Postan, M. M., and E. Miller, eds. *Economic History of Europe*, vol. 2. Cambridge: Cambridge University Press, 1952, 1987.

Powell, Elias T. *The Evolution of the Money Market (1385–1915): An Historical and Analytical Study of the Rise and Development of Finances as Centralised, Co-ordinated Force*. London: n.p., 1915.

Price, M. J., ed. *Coins*. New York: Methuen, 1980.

Reade, J. E. *Assyrian Sculpture*. London: British Museum, 1983.

Rich, E. E., and C. H. Wilson, eds. *Economic History of Europe*, vol. 4 and 5. Cambridge: Cambridge University Press, 1967, 1977.

Saggs, H. W. F. *Everyday Life in Babylonia and Assyria*. New York: G. P. Putnam and Sons, 1965.

St. John, J., ed. *Noble Metals.* Alexandria, Va.: Time-Life Publications, 1984.

Sherratt, A. *Cambridge Encyclopedia of Archaeology.* New York: Crown Publishers, 1980.

Singer, C., E. J. Holmyard, and A. R. Hall, eds. *A History of Technology,* vol. 1 and 2. Oxford: Oxford University Press, 1979.

Slater, R. *Portraits in Silicon.* Cambridge: MIT Press, 1992.

Smith, Lacey Baldwin. *The Horizon Book of the Elizabethan World.* Boston: Houghton Mifflin, 1967.

Thompson, J. G. *Mining and Metallurgy,* New York: American Institute of Mining and Metallurgical Engineers publication, May 1940.

Vandiver, P. D., O. Soffer, B. Klima, and J. Svoboda. *Science* 246 (1989).

Walker, C. B. F. *Reading the Past: Cuneiform.* Berkeley: University of California Press, 1987.

Wertime, T. A., and J. D. Muhly. *The Coming of the Age of Iron.* New Haven: Yale University Press, 1980.

Woolley, L., and P. R. S. Moorey. *Ur of the Chaldees.* Ithaca, N.Y.: Cornell University Press, 1982.

Yadin, Y. *Bar Kokhba.* London: Weidenfeld and Nicolson, 1971.

Zeleny, R. O., ed. *World Book Medical Encyclopedia.* Chicago: World Book Inc., 1988.

Zernick, Chloe. *Short History of Glass.* Corning, N.Y.: Corning Museum of Glass Press, 1980.

Index

age hardening, 195
aging temperature, 194
aircraft and aerospace industry, 3–4, 40, 48, 190, 195, 261, 262, 279–81
air disasters, 3–4, 7, 197, 217, 227
alchemy, 54, 72, 139, 183
alloys, 3, 40, 48, 59–63, 86–89, 143, 170, 194–95, 199, 252, 280
alloy steels, 213
alumina, 128, 181
aluminum, 40–41, 51, 97, 181–82, 186, 187–97, 198, 240, 252, 254, 269–70
 alloys, 3, 40, 48, 195, 261
 compounds, 186
aluminum-copper alloy, 195
aluminum oxide, 181, 252
amalgamation, 75
ammonia, 201–2
amorphous solids, 99
Anatolia, 8, 27, 28, 52, 89–90, 92, 108
animal horn and sinews, 18, 258–59
anisotropic strength, 262–63
annealing, of metals, 46–47
antimony, 143
antimony oxide, 117–19
antler, 18
Apollo space program, 273
aqua regia, 199, 200
Arabic numerals, 138–39, 266, 267
Aristotle, 182–83
arrows, 18, 19

arsenic, 61–63
artisans, 36–37, 38–39, 174
Asia, 17, 257–59. *See also* Far East; Mesopotamia
astronomy, 139
atmospheric pressure, 167
atoms:
 arrangement of, in crystals, 44
 bonding, in metals, 220. *See also* bonding, atomic
 constitution of, 187
austenite, 86–89
automobile industry, 212
axes, hand, 17–18

Bacon, Roger, 175
Backeland, Leo, 231
Bakelite, 230–31, 257
banking, 179
Bardeen, John, 270
Barnaby, Nathaniel, 210
batteries, 191–92, 193. *See also* electric battery
bauxite, 190
benzene rings, 236
Berthollet, Claude Louis, 205
Berzelius, Jöns Jakob, 185–86
Bessemer, Henry, 2, 206–8
Bessemer process, 2, 206–8, 209
billiard balls, 227–29
binary system, 266–68
bitumen, 17–18, 25, 126–27
Black, Joseph, 184
blast furnaces, 151, 165
bloom iron, 85
body-centered cubic structures (BCC), 86, 241, 254

Index

Boeing Co., 4, 197
bonding, atomic, 219–20, 270
bone, 7, 18, 19, 20, 251, 255–56
books, 141
boron, 252
  fibers, 260–61
bottles, glass, 115
Boulton, Matthew, 171
bow (weapon), 19, 257–59
brass, 83, 165, 170, 172
Brattain, Walter, 270
Brauer, Eberhard, 201
bricks, sunbaked mud, 25
bridge construction, 204, 212
Britain, 2, 4, 6, 152–55, 160–66,
  178, 206, 208, 213
brittleness, 47, 88, 122, 206, 210,
  240
bronze, 4, 49–67, 78, 86, 91–93,
  109
  cannons, 160. *See also* cannons
Bronze Age, 4–5, 59, 109
Brooklyn Bridge, 212
buildings:
  construction, 210–12
  materials, 124–33
Buna S. rubber, 225
Byzantine Empire, 136, 160

calcium oxide (lime), 103, 105,
  127–28
calx, 183, 184
camphor, 228
cannons, 155–61, 172. *See also*
  bronze; iron
capital, 178–79
capitalism, 151–55, 179
carats, 240–41
carbon, 21–22, 54–55, 86–89, 186,
  203, 205, 219–20, 239, 246–48,
  252, 253, 262

carburization, 87, 89
Carnegie, Andrew, 2
Carnot, Sadi, 174
casting, 63–66, 142–43. *See also*
  iron, cast
catalysts, 200–201, 233, 234, 235
catalytic converters, 201
Catholic countries, 179–80
Cavendish, Henry, 184
cellophane, 230
celluloid, 228–29
cellulose, 257
cellulose nitrate, 228–30
cellulose trinitrate, 228–29
cement, 128, 132
cementation, 205
cementite, 88
ceramic fibers, 264
ceramic oxides, 254
ceramics, 7, 8, 19–22, 27, 43,
  250–51, 280, 281
Chalcolithic Period, 23
chalk, 128
charcoal, 1, 53, 96, 157, 161, 162,
  164, 181, 239
  furnaces, 165
Chardonnet, Hilaire, 230
chariots, 36
chemical reactivity, 187
chemical warfare, 158–59
chemistry, 182, 185
China, 6, 136–37, 141, 144–45,
  150, 155–59, 165–66, 176–77
chips, silicon, 272–76
chromium, 213, 252
cinnabar, 75
circuits, integrated, 273–76
cities, birth of, 39–40
clay, 8–9, 19–37, 127–28, 144
  tablets, for writing, 33–35
coal, 54–55, 163–66, 170–71, 178

Index

mines, 155
coins, 76–81
coke, 54, 164, 165
cold-rolling of steel, 212
color, in glass, 117–20
combustion, 184–85
compass, 176
composite materials, 15, 40, 46, 55, 250–64, 280. *See also* metal-ceramic composites; polymer composites
compressive strength, 122, 257–59
computers, 265–76, 281–82
concrete, 15, 123, 125, 131, 162
condensation polymerization, 231
conductivity, 41, 268–70. *See also* electrical conductivity; thermal conductivity
Copernicus, 182
co-polymers, 237
copper, 28, 29, 39, 43–46, 49–67, 75–76, 79, 117, 181, 186, 240, 254
copper-arsenic alloys, 58, 59–63
ores, 61–62
copper-ore smelting, 51–54, 90–91, 96
copper oxide ore, 181
core-forming technique, 106, 107, 112, 115
corrosion, 213
covalent bonding, 220, 270
cristobalite, 101
crucible process, 205
Crusades, 159–60
cryolite, 191
crystals, 44–48
cupellation, 70–71, 76
cuprous oxide, 119
currency, silver and gold, 76–81

cylinders, 170–72
Cyprus, 90–91

Dacron, 230
Dalton, John, 185
Damascus sword, 137–38
Darby, Abraham, 164, 170
Dark Ages, 140
Davy, Humphry, 187, 188–89
decarburization, 205–6
deforestation, 162–63
devitrification, 101
diamonds, 22, 54, 72, 101, 220, 238–49
dislocations, 45–48, 87, 88, 193, 195, 272
dissolution, 193–94
divining rods (dowsers), 55–56
dopants, 270–72, 273

ebonite, 223–24
Edison, Thomas Alva, 193, 253
Egypt, 107, 108, 110, 112, 143
Eiffel, Gustave, 211–12
elastic deformation, 41–42, 45, 222
elastic energy, 257–58
elastomer, 226
electrical conductivity, 240
electric battery (voltaic pile), 188–89
electricity, 182, 186–89, 190–93
electric motors, 192
electrochemical industry, 189
electrolysis, 191–92
electrum, 70, 77
elements, classification of, 185–86
elevators, 211
Empedocles, 182
enargite, 61–62

energy:
elastic, 257–58
kinetic, 258
mechanical, 22
steam, 166, 178
engines, steam, 83, 167–74, 182,
203, 204, 209
ENIAC computer, 268, 276
erosion, of rocks, 15
Europe, 19, 20, 177
explosives, 201

face-centered cubic structures
(FCC), 44, 86, 241, 254, 255
faience, 105
failure, metal, 280
Fairchild Semiconductors, 272–73
Faraday, Michael, 192–93, 213,
219
Far East, 136–40, 177
fatigue, metal, 4, 7, 196–97
Fawcett, Eric, 232
ferrite, 86–89
fertilizers, 201
fiber axis, 262–63
fiberglass, 7, 259–60
fiber-optic cables, 100
fibers, 256–64
Fibonacci, Leonardo, 138–39
finance, large-scale, 179
fire lances, 157
fishhooks, 19
fishing rods, 260–61
flamethrowers, 156, 157
flax, in polymer matrix, 257
flint, 15, 18, 29
sickle blades, 24
tools, 17
Flussofen (flow oven), 151
flux, 91

food:
storage of, 8, 27, 32–33
supply of, and innovation, 16–17,
32
forging, of iron, 150. *See also* hot-
forging
fracture, 41
of glass, 122
of steel, 88
France, 154, 178
Franklin, Benjamin, 192
fuel efficiency, in cars, 280–81
furnaces, 113–14, 148–50, 151,
165. *See also* open-hearth
furnace process, blast furnaces
fused silica, 100

galena, 71
Galileo, 120, 167, 174–75,
179–80, 182
gangue, 91
gasoline, 127, 156, 174, 219
unleaded, 201
gears, 172
General Electric Research
Laboratories, 248
generators, electric, 192, 193
Germany, 2, 154, 208
Gibson, Reginald, 232
glass, 5–6, 7, 14, 98–123, 128,
247, 250
bottles, 115
manufacture, 162, 177
windows, 116
glassblowing, 97, 113–16, 177
glass transition, 226
glassy polymer, 226
glazes (ceramic), 105–6
gold, 6, 51, 68–81, 161–62, 177,
181, 200, 202

Index

currency, 76–81
extraction of, primitive techniques, 69
Goodyear, Charles and Nelson, 223
graceful failure, 280
grain boundaries, 47–48, 163, 272
grains, cultivation and harvesting of, 26–27
graphite, 7, 21–22, 54–55, 72, 239, 241, 246–48, 253–54, 260–61
graphite fiber-reinforced composites, 40, 55, 263–64
Greek fire, 156
Greeks, 6, 72–74, 108, 128
Gresham's law, 80
grinding utensils, 24
Guericke, Otto von, 167
guncotton, 228–29
gunpowder, 6, 135, 155–59
Gutenberg, Johann, 141, 143, 146
gypsum, 127–28

Hall, Charles, 190–91
Hall-Heroult electrolysis cell, 192
Hall-Heroult process, 191
hammering, hardening of metals by, 43–46
Hammurabi, 35–36
Hancock, Thomas, 224
handles, 17–18
hardening, 43–46, 60, 61, 87, 193. See also age hardening
hardness, 239–40, 255–56
harpoons, 19, 24
Harrison, John, 172
heat, and transformation of materials' properties, 22
heat-treating, of iron, 85

hematite, 84, 91
Heroult, Paul, 190–91
high-density polyethylene (HDPE), 233
High Speed Civil Transport, 279–80
high-sulfur hard rubber, 226
high-temperature furnaces, 113–14
high-temperature materials, 252–56
high-temperature superconductors, 9–10
Hochstetter (German engineer), 152–53
Hoerni, Jean, 273
Hoff, Ted, 275
Home Insurance Company (Chicago), 210
Homo, evolution of, 16–17
Hooke, Robert, 229–30
hooks, 24
Hoover, Herbert, 56
horn, animal, 18, 258–59
hot blast, 204
hot-forging, 88–89, 151
human history time scale chart, 12
Huntsman, Benjamin, 205
Huygens, Christian, 173
Hyatt, John and Isaiah, 228
hydraulic cement, 128–29, 132
hydrogen, 184

IBM Co., 276
I. G. Farbenfabriken Co., 225
incandescent lightbulb, 192
India, 136–39, 242–45
India ink, 141
Industrial Revolution, 83, 155, 165, 167, 178, 179, 182
information, storage of, 8, 32–35

Index

ink, 141, 143
integrated circuits, 273–76
Intel Co., 274, 276
intermetallic compounds, 254
internal combustion engine, 174
inventions, dynamics of emergence of, 172–75
ionic bonding, 220
ions, 118
iridium, 200
iron, 1–2, 4–5, 40, 96–97, 109, 115, 118, 148–49, 161, 163, 176, 178, 181, 203–5, 241, 248. *See also* bloom iron; pig iron
  alloys, 86–89
  cannons, 160–61
  cast, 86, 132, 150–51, 164–66, 170,172, 204, 208–9. *See also* casting
  mines, 155
  smelting of, 84–88, 90–91, 96, 148, 150, 162–63, 165, 178, 204
  wrought, 86, 89, 96, 150–51, 163, 165, 204, 208–9
Iron Age, 2, 4–5, 55, 82–97, 109
Ironbridge (Severn River bridge), 165, 208
iron oxide ore, 181
Islamic cultures, 135–40, 145, 148, 159–60, 177
isoprene, 225
isotactic structures, 234
isotropic strength, 263
Italy, 174–75, 177
ivory, 18, 227–29

Janety, Marc-Etienne, 199
Japan, 9, 20, 213
Jenney, William, 210–11
jet aircraft, 3–4, 262

jet engine turbine blades, 48, 252, 263–64
Jews, 4–5, 109–14, 134–35, 180
Judaean Desert Treasure, 58–59

Kelly, William, 207–8
Kepler, Johannes, 182
Kevlar, 217, 230, 260, 262
Kilby, Jack, 274–75
kilns, 53–54
kimberlite, 246
kinetic energy, 258
Klaus, Karl, 200
knives, 17
Korea, 142–43

La Condamine, Charles de, 218
Lagrange, Joseph-Louis, 185
latex, 218
Lavoisier, Antoine-Laurent, 75, 183–85, 199
lead, 29, 71–72, 75–76, 79, 143, 181, 201, 253
  ores, 70–71
  poisoning, 71–72
Leeuwenhoek, Antony van, 120
lenses, 120
Levallois technique, 17
levers, principle of, 169
lightbulb, incandescent, 192
lime, 103, 105, 127–28
  cement and mortar, 128–29
limestone, 27
linear unbranched polyethylene, 233
linseed oil, 143
liquation, 75–76, 152
literacy, 180
locomotives, 203
longbow, 158
lost wax technique, 58–59

Index

Lower Paleolithic period, 16–17
Lucite, 236

machine-driven economy, 178
Macintosh, Charles, 219
magnesium, 181, 186
magnesium oxide, 252, 254–55
magnetite, 84, 90
malachite, 27, 53
mammoth bones, 18
manganese, 207, 213
manganese oxide, 207
marble, 27
martensite, 88, 101, 194, 195, 210, 213
Martin Co., 209
marvering, 106
mass market, 164, 178
mass production, 65–66
materials:
  building, 124–33
  changing the shape of, 20
  discoveries by craftsmen and artisans, 181–82
  man-made, first, 20
  trade in, 27–28, 29, 31–32
matrices, 256–64
measurement instruments, 184
mechanical energy, 22
mechanization, 149
melting points, 100, 253, 270
memory chips, 274
mercury, 75
Mesopotamia, 8, 27, 29–37, 52, 94–95, 106–14, 126–28, 140, 143, 162
metal-ceramic composites, 264, 280
metallic alloys, 280
metallic impurities, in glass, 117–19

metallic matrices, 264
metallurgical industries, 151–55
metallurgical works (Saigerhütten), 152
metals, 7, 251–52, 268–70
  abundance of, in earth's crust, 50
  atomic bonding in, 220
  basic properties of, 38–48
  concentrations of, in ores, 50–51
  cracks in, vs. in glass, 120–22
  fatigue, 4, 7, 196–97
  vs. fiberglass, 260
  inception of human use of, 49–67
  modern, birth of, 176–202
  pure, 3, 43–48
  vs. rubber, 222
  strengthening of, 193–95
  strength of, 42–48
metal type, 142–43
metastable structures, 101–2, 247
metastyrene, 236
meteorites, as sources of iron, 83
microchips, 274
microphone, 193
microprocessor chips, 275–76
Middle Ages, 132–33
millefiori, 107
miniaturization, 281–82
mining, 55–58, 74–75, 151–55, 166, 169–72, 178
Minuteman missile program, 273
mixtures, rule of, 256
Modern Era, 22–23
molds (glass), 115
molybdenum, 254
monasteries, 149–50, 180
money, 76–81
Monge, Garspard, 205
mortar (building material), 126, 128–32
mortar (tool), 24

Index

Morveau, Guyon de, 199
mosaics, 107
motion picture camera, 193
Motorola Co., 276
movable type, 141–43, 146
mud brick:
  construction, 125–26
  with straw, 256–57
Mushet, Robert, 207

nails, 164
National Aerospace Plane (NASP),
  279
natron, 102–3, 105
Natta, Giulio, 234, 237
natural philosophers, 174, 182
natural resources, and national
  economies, 9
Near East, 4–5, 23–37, 69–70, 92,
  103–14, 136, 150, 177. See also
  Mesopotamia
needles, bone, 18
Neolithic Period, 19–20, 23, 28
net sinkers, 24
network modifiers, 103
Newcomen, Thomas, 169–72
Newton, Isaac, 182
nickel, 97, 181–82, 213, 233, 252
  alloys, 3, 48, 252, 254
niobium, 254
nitrate compounds, 201
nitric acid, 201
noble metals, 68, 198
nondirectional bonding, 220
Noyce, Robert, 273, 274, 275
nylon, 217, 230, 233, 262

obsidian, 14, 15, 27, 29, 102
Oersted, Hans Christian, 189
Ohm's law, 268–70
oil, 54–55, 163

open-hearth furnace process, 209
ores, 50–51, 181
osmium, 200
Ostwald, Wilhelm, 201
Otis, Elisha, 211
Ottoman Turks, 160, 259
oxidation-reduction reactions,
  51–52
oxides, 51
oxygen, 183, 184–85

palladium, 200
Pantheon, Rome, 131
paper, 6, 135, 140–46, 227–28
Papin, Denis, 167–69, 173
papyrus, 33, 143, 144
parchment, 144, 145
Parkes, Alexander, 228
Parkesine, 228
patent laws, 224
pearlite, 88
Persian culture, 6, 108, 136
Perspex, 236
petroleum, 200
pewter, 143
Phelan and Collender Co., 227
phlogiston, 183–85
phonograph, 193
phosphorus, 206
pig iron, 205, 206, 207
pistons, 168–69, 170–72
plastic deformation, 41, 42–47,
  163, 204, 222
plasticizers, 236
Platan, Baltazar von, 248
platina, 198
platinum, 186, 198–202, 230
platinum group, 200
platinum-rhodium alloy catalysts,
  201
Plexiglas, 236

300

Index

plows, 39
Plunkett, Roy, 235
polycrystalline solids, 47–48
polyethylene, 219, 221, 232–37, 262
polymer composites, 281
polymer matrices, 257, 259–62
polymers, 55, 215–37, 239, 281
polymethyl methacrylate (PMMA), 236
polypropylene, 234–35
polystyrene, 236
polyvinyl chloride (PVC), 236
population, density of, and building materials, 7
potash, 162
potassium, 186
potassium hydroxide, 189
potassium nitrate, 157
pottery wheel, 29
pozzulana, 128–29
precipitation, 48, 195
precipitation hardening, 48, 252
Priestley, Joseph, 183, 219
printing, 145–46
 ink, 143
 press, 140–46, 180
Protestant countries, 141, 179–81
pure metals, 3, 43–48
pyramids, 40
pyrite, 84

quartz, 27, 100–101
quenching, 88, 89, 93–94, 194

radar, 232
railroads, 173
rayon, 202, 217, 230
Réaumur, René Antoine de, 205
reduction, 181–82, 186, 189
Renaissance, 140

resins:
 phenolic, 231
 tree, 17–18
rhodium, 200
rockets, 157
rocks, 15
Roebling, John, 212
Roebuck, John, 171
Rolls-Royce Co., 263–64
Romans, 74–76, 96, 108, 116, 128–31, 132, 134–35, 144, 162, 177, 243
Rothschild Lycurgus Cup, 119–20
rubber, 7, 217, 218–19, 221–27
ruthenium, 200, 252

salt, 25
saturation, 194
Savery, Thomas, 169
Schönbein, Christian, 227–28
scientific method, 175, 182
scientists, 174
scrapers, 17
seashells, 29
semiconductors, 268–70
shape, altering of, 20–22
ships, 204
 hulls, 209, 210
Shockley, William, 270
sickles, 24, 26–27, 43
side groups, 220–21, 236, 237
Siemens Co., 209
silica, 99, 100–102, 105, 123, 128
silicon, 2–3, 8–9, 265–76
silicon dioxide, 273
silicon nitride, 252, 280
silk, 217, 229–30
Silk Road, 159
silver, 6, 51, 68–81, 93, 154, 161–62, 177, 181, 202
 coins, 76–81

Index

sinews, animal, 258–59
skin, as a polymer, 217
skyscrapers, 211–12
slaking, 128
slurry, 68
Smeaton, John, 132
smelting:
  with coal, 163
  of copper ores, 51–54, 90–91, 96
  of iron, 84–88, 90–91, 96, 148,
    150, 162–63, 165, 178, 204
  of silver, 70–71
  of sulfur-rich ores, 55
sodium, 186
sodium carbonate (natron), 102–3,
  105
sodium hydroxide, 189
sodium silicate glass, 103
solids, amorphous, 99
solid-solution hardening, 60, 61,
  87, 193
South America, 6, 198, 245–46
Spain, 6, 147–48, 161–62, 180
spear throwers, 19
Spiegeleisen, 207
sporting goods, 261
stained glass, 117
stannite, 62, 90
Staudinger, Hermann, 231
steam energy, 166, 178
steam engines, 83, 167–74, 182,
  203, 204, 209
steel, 1–2, 39–41, 83–84, 88,
  93– 97, 101, 125, 137, 203–14,
  247
  stainless, 213
Steel, Age of, 208
steric hindrance, 220, 234
stiffness, 41–42, 222
stone, 8, 20, 24, 162
Stone Age, 4, 13–37

straighteners (of wooden shafts),
  18
strength:
  anisotropic, 262–63
  compressive, 122, 257–59
  and crystal structure, 44–48
  isotropic, 263
  tensile, 122, 257–59
  theoretical and experimental, 45
stress, on metals, 42
Stückofen (wolf furnace), 148–49
styrene, 236
Styrofoam, 236
sulfur, 25, 157, 163, 223, 226
sulfur-rich ores, 55
superalloys, 252
superconductors, high-temperature,
  9–10
supersaturation, 194
synthetic fibers, 230

technology:
  incubation time, in ancient vs.
    modern societies, 113
  innovation in, related to defense
    industry, 274
Teflon, 217, 235
telegraph, 193
tempering, 88, 89, 122, 210
Tenant, Smithson, 200
tennis rackets, 40, 260–61
tensile strength, 122, 257–59
Texas Instruments Co., 274–75
thermal conductivity, 240
thermodynamics, 174, 246–48
thermoplastics, 229, 237
thrust-to-weight ratio, 281
tin, 59, 61–63, 79, 91–92, 109,
  143
tires (pneumatic), 224–25
titanium, 233

Index

tobomorite, 129
tools:
  metal vs. wood vs. stone, 65–67
  simple, 15–17
  toothed implements, 17
Torricelli, Evangelista, 167
trade:
  in materials, 27–28, 29, 31–32
  silver and gold as mediums of, 76–81
train tracks, 204
transistors, 268–74
tungsten, 213, 254
turquoise, 27, 56
type-metal, 142–43

unit cells (crystals), 44
United States, 2, 6, 206, 208, 213
United States Steel Corporation (USX), 2, 213
utensil money, 77

vacuums, 167–69, 172
vacuum tubes, 271
Vandermonde (French scientist), 205
viscose, 230
Volta, Alessandro, 188
vulcanization, 223, 226

Wang, Chen, 141–42
water-lifting apparatus, 169
waterpower, 148–50, 166, 178, 192, 204
Watson, Robert, 232
Watt, James, 171–73, 174
wheat, domestication of, 26–27
whiskers, 45–46
Wilkinson, John, 172, 204
wind power, 149, 166
Wollaston, William, 200

wood, 7, 20, 54, 162–64, 165, 181, 217, 251, 255, 258, 260
woodblock printing, 141, 145–46
wood pulp, 227–28
work, 168–69
work-hardening process, 43–46, 61, 193
written language, 32–36, 108, 140–43

yield stress, 42–48, 196
Young's modulus of elasticity, 41–42

zero, concept of, 139
Ziegler, Karl, 233, 237
ziggurats, 40, 126
zirconium, 233
zirconium oxide, 252, 255